Local Cells, Global Science

The book examines for the first time the extent of transnational movements of tissues, stem cells, and expertise, in the developing governance framework of India. It documents the impact of local and global governance frames on the everyday conduct of research, by tracing the journey of 'spare' human embryos in IVF clinics to public and private laboratories engaged in isolating stem cells for potential therapeutic application. The empirical research is informed by a comparative analysis of the ethical, religious and social issues in North America and Europe, and compared with other tissue and organ donations already prevalent in India. The discussion also examines the gender dimension as a potential site for exploitation in the sourcing of embryonic and other biogenic materials, and suggests that a moral economy has developed in which the ethical values of the global North support and encourage the donation of abundant and ethically 'neutral' embryos by the South.

Aditya Bharadwaj is a Lecturer in the School of Social and Political Studies at the University of Edinburgh, UK. His principal research interest is in the area of new reproductive, genetic and stem cell technologies and their rapid spread in diverse global locales ranging from South Asia to the United Kingdom.

Peter Glasner is Professorial Research Fellow for CESAGen at Cardiff University, UK. His long-standing interests are in the organization and management of the new genetics, the development of innovative health technologies, and in public participation in techno-scientific decision-making.

Genetics and Society

Series Editors: Paul Atkinson, Associate Director of CESAGen, Cardiff University; Ruth Chadwick, Director of CESAGen, Lancaster University; Peter Glasner, Professorial Research Fellow for CESAGen at Cardiff University; and Brian Wynne, member of the management team at CESAGen, Lancaster University

The books in this series, all based on original research, explore the social, economic and ethical consequences of the new genetic sciences. The series is based in the ESRC's Centre for Economic and Social Aspects of Genomics, the largest UK investment in social-science research on the implications of these innovations. With a mix of research monographs, edited collections, textbooks and a major new handbook, the series will be a major contribution to the social analysis of new agricultural and biomedical technologies.

Series titles include:

Governing the Transatlantic Conflict over Agricultural Biotechnology: Contending Coalitions, Trade Liberalisation and Standard Setting
Joseph Murphy and Les Levidow (2006)
978-0-415-37328-9

New Genetics, New Social Formations
Peter Glasner, Paul Atkinson and Helen Greenslade (2006)
978-0-415-39323-2

New Genetics, New Identities
Paul Atkinson, Peter Glasner and Helen Greenslade (2006)
978-0-415-39407-9

The GM Debate: Risk, Politics and Public Engagement
Tom Horlick-Jones, John Walls, Gene Rowe, Nick Pidgeon, Wouter Poortinga, Graham Murdock and Tim O'Riordan (2007) 978-0-415-39322-5

Growth Cultures: Life Sciences and Economic Development
Philip Cooke (2007) 978-0-415-39223-5

Human Cloning in the Media
Joan Haran, Jenny Kitzinger, Maureen McNeil and Kate O'Riordan (2007) 978-0-415-42236-9

Genetically Modified Crops on Trial: Opening Up Alternative Futures of Euro-Agriculture
Les Levidow (2007) 978-0-415-95541-6

Local Cells, Global Science: Embryonic Stem Cell Research in India
Aditya Bharadwaj and Peter Glasner (2008) 978-0-415-39609-7

Handbook of Genetics and Society
Paul Atkinson, Peter Glasner and Margaret Lock (2008) 978-0-415-41080-9

The Human Genome
Chamundeeswari Kuppuswamy (2008) 978-0-415-45857-3

Debating Human Genetics: Contemporary Issues in Public Policy and Ethics
Alexandra Plows (2008) 978-0-415-45109-3

Local Cells, Global Science

The rise of embryonic stem cell research in India

Aditya Bharadwaj and Peter Glasner

Routledge
Taylor & Francis Group

LONDON AND NEW YORK

First published 2009
by Routledge
2 Park Square, Milton Park, Abingdon, Oxon, OX14 4RN

Simultaneously published in the USA and Canada
by Routledge
711 Third Avenue, New York, NY 10017, USA

Routledge is an imprint of the Taylor & Francis Group, an informa business

Typeset in Times New Roman by
Taylor & Francis Books

British Library Cataloguing in Publication Data
A catalogue record for this book is available from the British Library

Library of Congress Cataloging in Publication Data
  Bharadwaj, Aditya, 1971-
  Local cells, global science : the rise of embryonic stem cell research in
India / Aditya Bharadwaj, Peter Glasner.
    p. ; cm. – (Genetics and society)
  Includes bibliographical references and index.
  1. Embryonic stem cells–Research–India. 2. Embryonic stem cells–
Research–Cross-cultural studies. I. Glasner, Peter E. II. Title. III. Series.
  [DNLM: 1. Embryonic Stem Cells–India. 2. Embryo Research–ethics–
India. 3. Embryo Research–legislation & jurisprudence–India. 4. Ethical
Relativism–India. 5. Internationality–India. 6. Social Change–India. QU
328 B575L 2008]
  QH588.S83B53 2008
  616′.02774072054–dc22

                                                              2008014922

ISBN13: 978-0-415-39609-7 (hbk)
ISBN13: 978-0-203-89103-2 (ebk)

Contents

Acknowledgements

The support of the ESRC is gratefully acknowledged. Part of this research was carried out under the aegis of the United Kingdom's Economic and Social Research Council (ESRC) funded Centre for Economic and Social Aspects of Genomics (CESAGen) and part at the School of Social and Political Studies, University of Edinburgh. In Cardiff we would like to thank Paul Atkinson, without whose kindness and support – particularly in the latter stages – this book would not have been completed, Helen Greenslade for her hard work and editorial support, and Andy Bartlett for his proof-reading assistance. At Routledge, we would like to thank our editor Ann Carter for her unstinting help and patience.

We would like to thank a number of family, friends and colleagues for their kindness, generosity and patient support. Special words of gratitude from Aditya are due to Marcia Inhorn for providing intellectual enrichment, many fruitful and stimulating collaborations and unwavering friendship. He remains indebted to his teacher and long-standing intellectual influence Veena Das. We are particularly grateful to Sarah Franklin, Janet Carsten, Nina Hallowell, Margaret Lock, Sheila Jasanoff, Elizabeth Roberts, Lawrence Cohen, Manuela Ciotti, Andrew Webster, Harry Rothman, Nikolas Rose, Jackie West, Rohit Barot, Sahra Gibbon, Carlos Novas, Daphna Carmeli, Carol Browner, Caroline Sargent and many, many others for numerous conversations and constructive criticisms. Special thanks are also due to Aditya's dear friends and colleagues Katie Featherstone and Liz Roberts for their fun company, uplifting support and steadfast camaraderie. Peter wishes to thank Jessica for keeping him grounded, and Angela for her patience and loving support in seeing this project through to completion.

In India, Aditya is truly indebted to the numerous individuals and couples who so openly and generously shared their life stories. He must also thank the clinicians, scientists and staff at various private research facilities, public hospitals and government departments in Delhi and Mumbai. Their invaluable assistance and cooperation made this research possible. For reasons of confidentiality they shall remain nameless but they know that he is truly grateful for their support for this research. Finally, Aditya must thank his family, especially his sisters and their partners, Meenakshi, Dilip, Ratna and

Sameer, for their loving and patient presence in his life. And last but not least, Ma and Bappi for always being there for him, for looking after him and nourishing him with love, conversation and delicious food!

Part of Chapter 2 was first published in Bharadwaj (2005b) and a different version of Chapter 3 appeared as Bharadwaj (2007). Different versions of parts of Chapter 6 appeared in Glasner (2005) and Glasner (2007).

1 The local, the global, and the contextual

An introduction

Introduction

The rise of embryonic stem cell research in India can be traced back to the middle of the twentieth century when a central role was assigned to science, technology and innovation as the principal drivers in transforming the economic and political fortunes of the nation–state. In the middle of the twentieth century two new state interests were added to the traditional one of national security, namely the promotion of science and economic development. Ashis Nandy argues that 'in the name of science and development one can today demand enormous sacrifices from, and inflict immense sufferings on, the ordinary citizen. That these are often willingly borne by the citizen is itself a part of the syndrome' (Nandy, 1996: 1). The notion of development as understood and taken for granted in global political and policy circles emerged in a speech given by US President Harry S. Truman in 1945; the idea of science as a state interest first appeared in a speech made by President John F. Kennedy. The speech declared one of America's major national goals to be the scientific feat of putting a man on the moon. In the past, Nandy argues, 'science was put to the use of the state but the state itself was never put to the use of science'. Kennedy's speech tied the very idea of science to the real and imagined goals of a modern nation–state.

Nandy's argument gains even more salience in the context of the burgeoning spread of neo-liberal governance modalities around the globe. The rise of the biotechnology of embryonic stem cells in India is, in part, symptomatic of this ongoing reordering and the emergence of a new global order. Contemporary India is rapidly challenging the long-established political and economic divisions between the North/South, First/Third Worlds, developed/developing economies and Western/Eastern cultures. The twentieth-century development discourse set in motion by Truman's speech, which envisaged unidirectional traffic from the global North to the local South, is in the process of being disassembled. In fact, the competition-driven market model of neo-liberalism requires an altered reading of Nandy's contention that in the name of science and development ordinary citizens often willingly bear enormous sacrifices and immense sufferings. As the book will show, in the *neo-liberal mode of production* a citizen's 'willingness' to bear suffering is distorted by socio-economic inequalities, just as the state's ability to demand sacrifice is enabled by the availability of gendered, stigmatized and impoverished citizens.

The precise significance of stem cell research in contemporary India will unfold in the course of this book, and no attempt is made to summarize it all here. Equally, the intention is not to provide a general introduction to the anthropology or sociology of stem cells and post-genomic biomedical knowledge. It will, however, be useful to offer an outline introduction to the subject-matter of this monograph.

Stem cells in general have, of course, been the subject of much popular attention and political intervention, in addition to social-scientific research and commentary. They are, in turn, aspects of a wider range of scientific and medical developments that have emerged in recent years and that have considerable economic, social and cultural significance (Brown and Webster, 2004). Stem cells are unspecialized, undifferentiated cells that have the capacity to develop into cells of different kinds of tissue or organs. Stem cells are, in other words, cells at a very early stage of their developmental life. They can be induced to develop into specific cell types, such as nerve cells or muscle cells.

There is nothing intrinsically controversial about stem cells. They occur naturally. They have also been used for therapeutic purposes for many years: bone-marrow transplants for the treatment of leukaemia depends on the capacity of stem cells in the bone marrow to turn into various lines of blood cells.

Stem cells have attracted a degree of controversy, however, because of the use of embryonic stem cells. Cells derived from the earliest stages of foetal development have the capacity to develop into any tissues and therefore provide scientists with the opportunity to grow cell-lines that can be used for research purposes, and to grow tissues (such as corneas and cartilage) that can be used in regenerative medicine. Human embryonic stem cells (hESCs) have been derived from 'spare' embryos created in the course of IVF treatment and not needed for implantation into the women's womb. Such embryos may then be donated by the couple who were receiving the IVF treatment for the purposes of scientific and medical research. Embryos may also be created through the use of cell nuclear replacement, where the nucleus from donated egg (oocyte) is replaced with the nucleus of a patient's cell. As we shall see, the availability of eggs and embryos has been of great significance in the emergence of India's stem cell research. (However, it is now being argued that the use of embryonic stem cells may eventually become obsolete, as it is now becoming possible to 're-programme' mature cells back to an undifferentiated stem cell state.)

This book exemplifies, with reference to the Indian context, that the creation, storage, circulation and use of stem cells are a matter of ethical and political debate. There are various national regimes of regulation, and differ from nation–state to nation–state. These differences in regulation create the possibility for international competition and markets in which ethical and legal regimes combine with economic and scientific capital in the creation of global markets in biomedical research.

Stem cell research is one major aspect of biomedical research, which is developing on a global basis. Increasingly national strategies in scientific investment are shaped by economic and cultural forces as never before (cf. Franklin, 2003b, 2005). National frameworks of regulation and state investment also interact with various private-sector interests. The latter include private biotech and pharma companies and research laboratories, and the private health care sector (such as IVF clinics). Local, regional and national configurations of these factors directly influence the level and direction of research effort in any national context.

This book, and the research on which it is based, are devoted to the particular case of India. Stem cell research is strongly driven by developments in the Asia-Pacific region, with major investments in China, Singapore and South Korea. India, which is part of this movement, is an especially interesting case, with a distinctive configuration of politics, economy and culture. As discussed in more detail at a later stage, the Indian government has made a major commitment to investment in scientific research and development. The supply of eggs and embryos on the other hand reflects the flourishing 'IVF scene' in India, which in turn reflects the high social value placed on fertility and reproduction. The favourable cultural environment is a reflection of the fact that the embryo is not accorded the kind of significance it is in Europe, Australia, the United States and other countries (Bharadwaj, 2005b). Fertility clinics are well established in India, and are lucrative sites of medical practice (cf. Bharadwaj, 2000, 2001). India has thus been especially well placed to develop a stem cell research effort, based on a combination of economic, ethical and cultural factors. These developments take place in a global context, and it is the aim of this book to explore the relationships between the local and the global in the circuits of stem cell research.

In examining contemporary developments in the generation, application and legislation of embryonic stem cells in India the cultural, social, economic and political processes that have an impact on those developments are analysed. In so doing, the book explores how these processes shape, and are shaped by, developments in other locales around the globe. The picture is at best partial, as the field is changing rapidly – locally and globally: the book captures a snapshot of a complex story. However, the analysis does go beyond the particular ethnographic exploration of Indian stem cell science, in order to address more generic analytic issues. The main thrust of the book is to explore what 'local' and 'global' might look like in a world where 'locales' for the production of new scientific knowledge and its application are departing from 'normative site specificity'. Former 'Third World' countries like India are reinventing themselves as global players, not merely economically, but also ethically and politically. They now participate, more than ever before, in re-framing national and international regulatory and ethical regimes. This recasting of the global is both intimately local and strategic. In other words, its unique local cultural and moral context gives India the edge in recasting herself as a global player in emerging scientific fields such as embryonic manipulation for stem cell extraction. At the same time, however, the dominant global patterns of neo-liberal statecraft and political economy put in place the material and cultural conditions under which a rise to global prominence might be achieved. The local circumstances that make possible rapid biomedical development must, therefore, be tempered to global ethical and political frameworks. Hence, the book seeks to explain the rise of embryonic stem cell research in India by arguing that a 'neo', as opposed to 'new', India is on the rise. This rise is achieved, in part, due to a number of structural, cultural, ideological, and material conditions, both local and global.

The book conceptualizes shifts in these local and global formulations through which India is being re-imagined and recast in the new century. In order to examine these transformations, a number of key concepts are deployed including 'dis-locations', 'moral economy', 'bio-crossings' and the 'liminal third'. These concepts are unpacked and related to the main theoretical frame of the local and the global in this introductory chapter and throughout the rest of the book. However, it is important to note that, as with any process of change, there are inherent tensions that India is encountering on her 'upward journey'. These creative tensions become visible as the country is steered into a neo-liberal political economy, towards a synthesis that is global and yet localizing, perhaps even uniquely 'Indianizing'.

There are obstacles to this process of transformation. They include entrenched structural constraints (and many do exist), and imagined traditions and their inevitable incompatibility and clash with supposed modernities. In addition, there is resistance articulated by actors who inhabit the institutions and endorse the ethical frameworks that are threatened as the local confronts the global. It is, moreover, a mistake to assume that what is happening in India is in some fundamental way unprecedented. A cursory glance at the history of India is enough to demonstrate that Indian history contains a series of cultural encounters that have had profound reconfiguring impacts on its culture and geographical spread. India is now negotiating the global, fast moving currents, on which circulate economic and biological capital, natural and human resources, 'local moral worlds' and global ethical and governance modalities. It is not the first time that India has confronted global economic and cultural transformations on an equivalent scale.

A brief overview of the way in which the notions of 'local' and 'global' are framed is presented in this chapter. This is followed by a brief introduction to the ethnographic fieldwork and its conduct. The chapter concludes with an overview of the subsequent chapters. This sets the context for further exploration of the various substantive themes within this book. As with any text, there are multiple ways to enter and exit this book. Each chapter can be read on its own, or in conjunction with one or more of the others. However, the book is best understood if it is read in its entirety in order to follow the conceptual threads and to understand the way in which they are woven into the ethnographic illustrations. This brings real-life context to conceptions about the place of science in a rapidly altering world where notions of the local and the global are being shaken up in an unprecedented fashion. Different readers will read the book in different ways, and will understand, critique and extend the formulations in their own way. It is the act of reading and not necessarily the act of writing that eventually creates a text.

Contextualizing the local and the global

More than ever, the local and the global are vantage points, as opposed to mere spaces, sites, and locales. That is, ideas of the local and global inform perspectives on selves and others. These can range from the self-description or self-imagination of a nation–state (as a politically defined and geographically bound space) through to cultures, individuals, and communities, through to economic and commercially driven notions of what a global site and its dynamic and shifting relationship to localities might look like. In the volatile world of commerce, trade, and international business – a world driven by venture capital, heavy state subsidies, investments, and the circulation of capital to sites deemed advantageous and 'safe investment havens' – the local and the global are repositioned, redefined and strategically reformulated. In other words, the contingent nature of locality is a complex of power and socio-political relations that move responsively to the centripetal forces of global economic production, circulation and consumption. The local and the global are no longer simply static, bounded sites that can be 'journeyed' to and from. In the high-stakes world of commerce, science and innovation, for example, we can no longer think of a global/local site being geographically bound, with policed national borders. Ideas, resources and capital increasingly journey, and, in so doing, they influence, have impacts on and shape national and international conceptions about politics, economics, and moral and ethical thinking. Whyte and Ingstad (2007) have recently argued that, rather than examining the global as the given circumstances in which the local exists, one might more fruitfully think of it 'in terms of the verb: "to contextualize", to weave together' (ibid.: 9). They propose a two-pronged approach to problematizing the links between local and global worlds. First, one must pay attention to the ways in which people imagine and mobilize the notion of the global in relation to particular projects that they are engaged in. An example would be calling upon a universal set of values and ideals such as the UN Declaration of Human Rights. Second, one must trace the 'movement of people, things, images, ideas, from one local world to another', thereby understanding the flow of phenomena, the 'channels' through which the move is accomplished, and the types of changes in the landscape such flows and moves bring about (ibid.: 10).

In a similar vein, Appadurai (1996) has engaged with the increasingly dynamic local flows that reconstitute the global in a multitude of ways using the metaphor of landscapes. Appadurai proposes an elementary framework for exploring the 'new global cultural economy' by examining the relationships between five dimensions of global cultural flows, which he terms; '(a) ethoscapes, (b) mediascapes, (c) technoscapes, (d) financescapes, and (e) ideoscapes' (ibid.: 33). He argues that the suffix-scape allows an appreciation of the fluid, irregular shapes of these landscapes, 'shapes that characterize international capital as deeply as they do international clothing styles'. Appadurai defines these landscapes as the building blocks of an 'imagined world', that is, multiple worlds that are constituted by the historically situated imaginations of people and groups spread around the globe. He argues that many people live in such imagined worlds, and are on occasion able to contest or even subvert the imagined world of the official mind and that of the entrepreneurial mentality.

In many respects, this book is an illustration of the imagined worlds inhabited by individual scientists, groups of patients, embryo donors and their families. It explores too the imagined worlds of the Indian and Euro-American states – their modes of governance, legal codes and ethical models. The imagined worlds extend to the local and global mass-media narratives of 'maverick science' and 'miraculous embryonic stem cells'. Many of these actors and institutions involved in these processes have conflicting interests. Others strive to find a common ground, melding together the local and the global ideological and normative forces which shape the contingencies that allow calculated investments in ideas such as science, regulation, ethics, morality and economics. These changes, broadly conceptualized as being symptomatic of an emergent global order, are producing 'dis-locations' that both define and realign the context of the local and the global. This is the theme of Chapter 2. Dis-location, in this respect, is an ethical, moral, geopolitical and economic digression from an imagined normative global order. It is a departure from a comforting state of affairs, and an arrival at the uncertainty of upheaval. Dis-location, in this sense, invokes a nostalgic view of past certainties and the tantalizing promise of an unpredictable future.

India is analysed as a dis-location on three levels. First, in a departure from its twentieth-century post-colonial incarnation as a Third World developing country, India is emerging as a global player, thus dis-locating the taken-for-granted, oppositional view of the post-war world as developed or developing. Second, India is rapidly challenging its established global image of a provider of call centres, cheap technical labour and information technology products, by aggressively colonizing the unlikely global site of biotechnology research and innovation. Third, India is a dis-location as it is routinely 'othered' as a locale involved in unethical and/or maverick science. It is often argued, in domains such as media and policy circles, that India is yet to achieve the 'high ethical and moral standards' that frame biotechnological research and application in the predominantly Euro-American global locales. However, as a nation–state, India is making concerted efforts to put in place governance and legislative protocols that resonate with the prevailing neo-liberal political and economic global climate (hereafter *neo*-Global).

These moves, that increasingly steer the booming political economy of a former 'Third World' nation, are conceptualized in Chapter 2 as the rise of a 'moral economy'. The question posed is as follows: What are the consequences when 'morality' is incorporated as a key component in the economic and political calculations of nation–states? It is suggested that when morality is brought into play in the service of producing economically viable entities, it produces much more than mere commodities. Under the neo-liberal order, the moral, in economic and political calculations, achieves a sanitizing effect for both the production and the consumption of entities by marking them as ethically untainted. This facilitates guilt-free circulation, consumption and accumulation. A neo-liberal 'moral economy' of this kind is one of the key features of a 'dis-locating' global landscape.

Wherever there are formidable structures, there is agency. The analytical challenge is to theorize and empirically elaborate without essentializing the 'brutalizing' effects of structure and the 'heroic resistance' of agents. In the context of the global spread of new reproductive technologies to 'third world' sites such as Egypt, Marcia Inhorn has argued that:

> [M]any medical anthropologists, myself included, who work in Third World settings have been guilty of 'romanticising' reproductive agency at the expense of constraint. In our efforts to show that 'subordinated people are not unreflecting automatons', but rather actors who demonstrate resolve, struggle, and problem solving in their quests to control their reproductive lives, we have sometimes overlooked, or at least underemphasised, the severe constraints that effectively limit reproductive choices, including the often painful decisions about whether and how to use reproductive technologies.
>
> (Inhorn, 2003: 17)

A 'bio-crossing' is a conceptual space where structure and agency can be unpacked in their proper context (Chapter 3). A bio-crossing is conceptualized as a passage or crossing that traverses the borders of biology, moves between biology and machine, and passes across geo-political, commercial, ethical and moral borders. These crossings are not always governed, inspected and evaluated. On the contrary, they may be illicit, unauthorized, subversive and rebellious. To undertake such a crossing is to enter a space that can exist independently of local and global normative prescriptions, from national border controls, to the official regulation of tissues and bio-genetic materials such as embryos, oocytes and blood. These crossings are seldom made through zones of ethical and moral consensus, but rather through zones of 'in-distinction', characterized by competing and contested ethical frames. Such in-distinction challenges the stability of norms and ideologies in any local/global formation. Agency in such bio-crossings is often severely curtailed by structures that police, patrol, and protect the dominant norms and ideas, underscoring the established social contract between the governing and the governed. On occasion, this may take the shape of open confrontation, a rejection of the dominant paradigm, or it may take the shape of negotiations, articulated from within the normative bounds of a stated ethico-legal or socio-political framework.

Chapters 3–6 explore the increasing resistance to proposed governance models that will shape the future direction of research in embryonic manipulation and application of stem cells in India. The ongoing negotiations between the various stakeholders are similarly examined. Marcia Inhorn (2003) has persuasively shown that, when transferred to global locales (such as Egypt) and detached from their points of origin and context of production, the assumption held by the global producer nations remains that the new reproductive technologies (or biotechnologies in general) are 'value-free, inherently beneficial medical technologies', and that these remain 'immune to culture and can thus be appropriately transferred and implemented anywhere and everywhere'. This assumption, according to Inhorn, 'is subject to challenge once local formulations, perceptions, and actual consumption are taken into consideration' (ibid.: 15).

There is an added dimension in the emerging global context of stem cell research, in which local governance and legal models are either expected or enjoined to meld with the ethical, moral and legal sensibility of the 'dominant global' producer nations. In a case, such as that of India, of economies emerging on to the *neo*-Global stage, such an ethical and philosophical merger is desirable, even expedient. This merger creates a level playing field and new markets where products and services that match 'global standards' can be circulated in line with the prevailing 'moral economic' sensibility. That is, a moral economy that demands – more than ever before – that moral and ethical concerns are 'built into' products as standard. Whose ethical and moral sensibilities these might be is, as yet, not open to negotiation, as the Euro-American global locales continue to lead the field on these critical philosophical matters. However, if the predicted rise of India and China is to become a reality in the twenty-first century, then an ethical dis-location that compels a 'culturally informed' rethink can be expected on the biotechnological horizon. For now, however, bio-crossings remain quintessentially 'anti-structural'.

To follow the fourth main conceptual (re)formulation, bio-crossings take place between the local and the global, in a 'liminal third space' (Chapter 5), a conflation of the existing concepts of 'liminal time' and 'third spaces' (cf. Van Gennep, 1960; Turner, 1967; Soja, 1996). When combined, liminal times and third spaces depart radically from the rigid, the predefined, and – most importantly – the opposed categories. In other words, the liminal third is a departure from a rigid script. It is an improvisation. The permutations and combinations are limitless, as are the consequences and outcomes. Liminal conditions are normally described in terms of temporary disruption and separation, followed by an eventual return to normal states of affairs. In contrast, a third space remains liminally open-ended. The liminal third, therefore, is a space that is anomalous, that is unpredictable. It provides an open-ended opportunity to critique, challenge and change states of affairs. The life and work of stem cell scientists within these unstructured liminal third spaces in India, and indeed the biological biographies of individual patients, are best understood when examined as part of such third spaces where bio-crossings take place. These spaces are both a challenge to the official governance models and to the moves to reformulate the ethical and moral controls on science within India. This process is in line with the prevailing canons of structured and rule-ordained practice, underpinned by philosophical principles that demand global bioethics in place of local, culturally-bound practice. These views are most graphically expressed through local and global media channels. They create scripts that construct polarities – good and bad science, scientists and mavericks, global and local – that actors in a liminal third space seek to subvert. It is ironic that the very act of documenting and narrating this complex story of local and global spreads creates methodological 'third spaces' with important representational consequences. It is this particular third space that we must now enter.

Ethnography in and of a third space

The study on which this book is based is an amalgam of anthropological and sociological approaches to ethnographic narration (cf. Atkinson, 1990). Anthropology and sociology have long and well-documented histories of ethnographic immersion in the local moral worlds of hosts or informants. However, since the late twentieth century sociological and anthropological ethnography has undergone profound change. Formerly taken-for-granted certainties concerning fieldwork and its products have been called into question; sociology and anthropology have followed parallel paths of reflexive critique in this respect. Michael J. Fischer recently described anthropology as a third space (Fischer, 2003). He locates the formation of the discipline in a space between the colonial desire for empire and the defence of the marginal, oppressed, subaltern and perhaps even the anomalous. This third anthropological space is engaged in evolving new, multicultural ethics and is busy making 'visible the differences of interests, access, power, needs, desire and philosophical perspective' between social actors and groups (ibid.: 3–8). Perhaps anthropology could have occupied this third space much earlier, had not its rhetoric of persuasion been too bound up with colonial administrative demands for static cultures. However, since the 'crisis of representation' occasioned by the publication of *Writing Culture* (Clifford and Marcus, 1986), and with the emergence of the 'ethnography in/of the world system' (Marcus, 1995), the methodological tool-kit of the discipline has altered beyond recognition. The accompanying rise of the, now problematic, indigenous ethnographer (Fahim, 1982; cf. Clifford, 1997), and the often diasporic identities of such ethnographers have contributed to the creation of third spaces from which to gaze, reflect, record and perform critique.

In this respect, *Local Cells, Global Science* is an account from a diasporic third space, conceived within the third space between anthropology and sociology, and narrates a partial tale of the emerging third space within which biotechnologies of embryonic stem cells have begun to flourish in India.

Until recently, the nation–state of India occupied the third space in the global hierarchy of nations. Post-independence India, through its chairmanship of the non-alignment movement, sought to locate itself in a third space between the Soviet bloc and the capitalist world led by the United States of America. Though seen as lying within the Soviet sphere of influence, India sought to remain entrenched in the non-aligned third space by opting for a mixed economy model that, at least in theory, favoured an eclectic blend of the socialist and capitalist economic modalities. The story of India's move from this 'third world' space to that of the 'developing' one, with aspirations to become a 'developed nation' in the new century, is also in part the story of the post-colonial rise of science and technology in the country (Bharadwaj, 2005b). Since the 'liberalization' of the economy in the 1990s and partial opening up of the Indian economy to the rest of the world, a new era of foreign direct investment and the rise of home-grown multinational corporations has dawned. India, we are told, has gone global.

The conceptual and theoretical thinking behind this book is a result of immersion in a multi-sited ethnography examining cultural complexities inherent in the production of embryonic stem cells in India. Emily Martin, in her research on perceptions of immunity in American culture (Martin, 1994), drew on Clifford Geertz's (1983) image of 'culture as an old city surrounded by modern suburbs', in turn drawing on an earlier analogy developed by Wittgenstein (1953) to describe language. Martin argued that this image of culture suggests that cultures can 'simultaneously involve structures built in different time periods and that inhabitants of those structures might be able to move about the landscape' (Martin, 1994: 8). She complicated this picture by adding a significant new dimension to this imagery, that of the electronic media, which could transport 'images, words, and ideas between the old city and the new suburbs, the rest of the country, and the world' (ibid.: 8). Martin's methodological construct typifies a truly multi-sited ethnographic practice, where the subject matter is dispersed in multiple locales/domains, and the subject of research is painstakingly traced through such disparate locales.

According to George Marcus, a multi-sited ethnography is an exercise in 'mapping the terrain', or rather the mapping of part of the terrain, as mapping, by its very nature, can never be total. Its goal is not 'holistic representation'. Rather, such a mode of ethnographic immersion in a cultural formation in the world system is an ethnography of the system, and therefore cannot be understood as a single site (Marcus, 1995: 99). For ethnography of this kind, therefore, there is no global in the frequently evoked local/global contrast, but rather the global is 'an emergent dimension of arguing about the connection among sites in a multi-sited ethnography' (ibid.: 99). Marcus further argues that when 'the thing traced' is within the 'realm of discourse', then the 'circulation of signs, symbols and metaphors guides the design of ethnography' (ibid.: 108).

This mode of doing multi-sited ethnography involves tracing the social correlates and groundings of associations that are most visible and alive in language use and print or visual media. The extracts from transcribed interview accounts, the official printed documents of regulatory agencies, and various globally situated media locales, become specific sites that feed and sustain the productions of ethnographic narration. For the present research, such an approach acquires even more importance, as the movement of images, words, and ideas between the 'old city', the 'new suburbs', between national and global entities, allows an empirical investigation of the circulation of ideas about embryos and tissues, as well as allowing tracing of the local and global material production of human embryos and embryonic stem cells. In probing the old (city) and the new (suburb), and their substantive relations with similar global locales, the research produces a complex picture. The mass media in this respect are a 'site' and a 'locale' that compel the work of imagination and inform the flow of communication in other, more dispersed, local and global sites. Therefore, the multiple locales that this research uncovers helps examine and problematize the supply of human embryos from the point of conception (or rather donation in IVF laboratories), through their eventual and ongoing manipulation in the public and private sector research facilities, and to their moral and ethical policing in local and global contexts.

The ethnographer 'other'

The experience of fieldwork undertaken in India by Bharadwaj, in large part, firmly located him in the third space. Srinivas once described himself as an 'oddball among anthropologists', since all of his fieldwork was carried out in his own country – India (Srinivas, 1997). The observation is interesting but at odds with the contemporary reality, in which increasingly rich insights into cultures come from a growing tribe of indigenous and native ethnographers (Fahim, 1982; Ohnuki-Tierney, 1984). However, Srinivas's contention (Srinivas, 1997: 22), that to study one's own culture is like studying the 'self-in-the-other and not a total other', captures some aspects of Bharadwaj's fieldwork experience. The category of 'other' is, in fact, never fixed and its meaning shifts according to context (Thapar-Björkert, 1999). Hence, while Bharadwaj was an insider, to the extent that he was born in India and could identify with his respondents on that level, he nevertheless, on occasions, faced the problem of being positioned as the non-resident

Indian (NRI) 'other' by his respondents because he was an academic based in Britain. NRI is a term used in India to 'other' people of Indian origin as someone permanently based in a foreign land. As Bharadwaj was an NRI to his respondents – infertile couples or patients, clinicians and scientists alike – he was often told, teased or warned that he could 'never understand them completely', that he was used to a 'different way of being', and that he was now perhaps too 'Westernized to understand how India worked'. The identity and authority of 'indigenous ethnographer' are compromised in such a context, thrusting the ethnographer into a third space of diasporic location and locution.

This, then, is the 'anthropological paradox', where, on the one hand, it is maintained 'that it takes one to know one, and on the other, that the hardest thing to know is one's own culture' (Napier, 2004: 860). However, as Clifford (1997) shows, the opposition between the native and the non-native, insider and outsider, is rightly being questioned despite fieldwork being 'rerouted by indigenous, postcolonial, diasporic, border, minority, activist and community based scholars' (ibid.: 77). According to Clifford, 'there is no undivided "native" position'. Drawing on Narayan (1993), he points out that native anthropologists and researchers, like all anthropologists and researchers, belong to several communities simultaneously. These multiple locations were brought home on a number of occasions in this project during formal and informal interactions, as the research interlocutors variously ascribed to, and withheld from, the ethnographer native advantage, authority and knowledge.

The empirical, the existential, and the experiential

The bulk of the fieldwork, which Bharadwaj undertook from 2004 to 2006, entailed observations in a stem cell research laboratory in New Delhi and extensive interviews with scientists engaged in isolating embryonic stem cells in Mumbai (formerly Bombay) and Delhi. Various government officials based in the Indian Council of Medical research and the Department of Biotechnology were interviewed. In addition, over forty semi-structured interviews were conducted with current and potential embryo donors in two IVF clinics in New Delhi. The first of these clinics is located in an upmarket South Delhi private hospital, with a middle- to upper-middle-class urban clientele, while the second facility is publicly-funded, based in New Delhi's Army Research and Referral Hospital (ARRH). The bulk of IVF treatment-seekers at the latter clinic are soldiers of the Indian army with rural, working-class backgrounds. Since Bharadwaj began studying the proliferation of IVF in India 10 years ago, the technologies of procreation have become available in the public sector. The class disparity between the clinics therefore provides a rare comparative insight, as assisted reproductive technologies in India continue to be a predominantly urban and middle-class phenomenon.

Fieldwork was also carried out in the Delhi clinic of Dr Geeta Shroff. Geeta Shroff has gained considerable fame in some quarters and notoriety in others. She claims remarkable therapeutic success in the use of stem cells for a variety of major medical conditions, but works independently of mainstream research and practice. Her work has been widely reported in the mass media. By gaining access to this clinic the research was able to document at first hand the working of a controversial location of stem cell 'experimentation' in India.

All interviews were tape-recorded, transcribed and thematically coded for analysis. The interviews with government officials, clinicians and research scientists were carried out in English. Follow-up interviews were recorded with a senior Indian Council of Medical Research official, and with Delhi-based stem cell clinician, Dr Geeta Shroff. Interviews with the army personnel and their spouses were conducted in Hindi, and later translated and transcribed by Bharadwaj. All the middle-class interviews in the South Delhi IVF clinic were conducted in English, with the notable exception of one conducted in Hindi. All interviewees were anonymized and given pseudonyms in order to protect their identities. However, with her express permission no effort was made to conceal the identity of Dr Shroff, as both she, and her clinic, were a central focus of the ethnographic component of the research. In addition, the global media interest in her practice has turned her into a public figure too controversial for anonymizing to succeed. The research also relied on extensive archival searches, print and news media analysis, and a detailed review of current and proposed Department of Biotechnology and Indian Council of Medical Research Guidelines and ethical codes of practice.

The economic rise of India was self-evidently present and observable in Bharadwaj's everyday movements through the city of Delhi and the financial capital Mumbai. The media hype was strongly corroborated on a daily basis as Bharadwaj negotiated the cityscapes. Chance encounters and conversations with ordinary citizens similarly revealed an upbeat and expectant outlook. For a certain segment of India, the imagined local and global were rapidly swapping positions. Excerpts from Bharadwaj's fieldwork diary capture the sense of collapsing domains of the local and the global in the urban centres of India. These musings also locate the ethnographic gaze and voice in the third space of an 'insider outsider'.

The accompanying representational dilemma remains an implicit corollary of occupying such an anomalous space. Crucially, it provides an insight into the persistent and entrenched social cleavages, accentuated by neo-liberal expansion, the slow pace of urban regeneration, and the accompanying explosion of consumption among the rapidly globalizing middle classes in India:

[T]hen there is the everyday living and travelling in Delhi. It really did surprise me how living in a 'developed' country sanitises the gaze. I am a lot more sensitive to Delhi's confounding, mind-boggling contradictions and diversity than ever before. It's hot, humid and the monsoon has finally descended on the city with all its fury. It is interesting how quickly one unlearns to appreciate rain living in the UK. Sometimes it rains and rains and then the day turns cloudy bringing much needed respite from the sweltering heat and humidity. There is a wafting smell of earth in the air, and everything looks washed and green. The city comes alive and with it sewers, piles of unclaimed garbage, pot holes and tadpoles! Cows saunter to the middle of the roads in search of rushing, manic traffic that blows away swarms of flies from their holy perch. And then as I stand before this hurly burly of a typical monsoon day – mud, muck, rubble, rubbish – I spot the new Tag, Dior and Tommy Hilfiger stores. The heaving, humid city recedes in the background, for the hard-working, globe-trotting, great Indian middle-class consumption is at long last in sight! From beyond the uninterrupted sea of soggy shanty towns rise glittering buildings, mega shopping malls, glass steel and yet more glass! Shining, floor upon floor of retail therapy and a sanctuary for the well travelled and the well (high) heeled! Beggars, lepers and the maimed extract their daily wage from the never-ending convoy of smart cars as they pause at traffic intersections. Little urchins sell flowers, incense and boxes of serviettes; some beg. People roll down their windows, just enough for the money to pass through the slit, momentarily letting in the fuming heat into the sanitised air conditioned comfort of their sedan, saloon, four-wheel drive. The light turns green and the restless, the guilty (the middle-class ethnographer included) flee, escape to another day, leaving behind dust, exhaust fumes, chocolate wrappers and a few million poor. I have to keep reminding myself what Mark Tully, BBC's man in Delhi once said in response to the question: 'How do you deal with poverty in India?' he simply replied, 'I don't have to. The poor do!'

(July, 2004)

In the run-up to the celebrations of sixty years of Indian independence, a lot of global media attention was focused on the 'spectacular rise of India' while acknowledging that over 300 million Indian citizens lived on less than a dollar a day. A disturbing statistic to many, but, to the ruling elite, mesmerized by the neo-liberal modality of substituting governance with technical fixing and troubleshooting, it was an impressive achievement that so much could be achieved in so little time. In other words, in a country of over a billion people, and an economy predicted to grow between 10–12 per cent a year, the hope remains that it is only a matter of time before the teeming millions will benefit from the fruits of globalization.

The interviews and participant observation were carried out over four months, from June to August 2004 and November 2004 to January 2005. The ethnographic location in Dr Geeta Shroff's Delhi clinic was by far the most complex space to negotiate. The first meeting in 2004 left Bharadwaj with an impression little different from that presented by familiar media reports that cast doubt on Dr Shroff's claims of success. However, unrestricted access to the clinic and the patients meant that, over a period of time, 'biological biographies' of individual patients began to surface, and what previously seemed like the untenable claims of a 'maverick clinician' began to take shape as the real-life experiences of individuals with long-documented social and medical histories of suffering.

The representational challenge of describing the dramatic admissions of healing, sensation in paralyzed limbs, and the ever-increasing flood of new arrivals from around the world in search of stem cells remained. The placebo effect was an obvious answer in many cases, but, equally, from 2004 to 2006, with every return trip to India, the same individuals continued to register improvement. How long could a mere placebo effect last? Unlike the international media and experts scattered around global locales, the decision to focus on the emerging and intertwining discourses about science, healing, ethics, regulation, economic and political interests seemed to be the best way forward. The complex story of the local and global spread of embryonic stem cells is not so much about truth claims, and whether these are credible or not, but rather it is about the kinds of discursive formations that gained power, credibility, and the way in which a complex of political and economic considerations drove the international, 'translocal', institution of what was now to count as truthful and credible, ethically and morally viable, or politically and economically expedient. The interactions with the embryo donors in the private clinic was an unsettling experience, as was their enrolment in the research through the bureaucratic articulation of ethical guidelines that ensured 'informed consent'. In many instances the couples were so preoccupied with the distressing and socially debilitating effects of infertility that they did not care what informed consent meant. On many occasions women broke down in tears during the course of the interviews, and on other occasions promised embryos in return for a quick resolution to their childlessness. Encounters in the army hospital were, however, different, as none of the soldiers interviewed were, as yet, in an active donor programme. Their idealism and sense of duty became an extension of their professional remit. Their will to sacrifice for the greater good was a far cry from the global ethical demands that are routinely enshrined in the moral calculus driving the political economy of neo-liberal nation–states. These encounters raised a number of similarly troubling and unresolved issues, some of which are tackled in the pages of this book, while others are in need of long and sustained reflection.

Bharadwaj's fieldwork in 2004 and 2005 was interspersed with periods of archival and media searches, transcription and coding with Glasner in the United Kingdom. This spatial and temporal dispersion of the research subject-matter created a third ethnographic space of the kind that could no longer be analysed in terms of the (first space) global and the (second space) local. The distinction became untenable as the ethnography moved through cyber-space, the ambivalent public/private space occupied by the research subjects in IVF clinics and their cryogenically preserved, cell-yielding embryos, to many international crossings and impromptu 'stem cell conversations' with the 'diasporic cargo' on cramped (third) spaces of Boeing 747s. Here we find another ethnographic paradox. On the one hand, as Veena Das (2000) concludes in the context of her research on organ transplantation, this mode of doing ethnography lacks the security of a locality to which the anthropologist can go, leaving his or her normal academic world behind. On the other hand, such a mode opens a terrain of analysis, cultural critique and ethical plateaus in an anthropological third space (Fischer, 2003).

The book: structure and content

The book is structured into five further substantive chapters and a concluding discussion. Each chapter is conceptually grounded and engages with a wide range of theoretical formulations to contextualize the ethnographic observations and empirical illustrations.

Chapter 2 proposes the concept of dis-location. It is argued that the rise of India on the global stage is symptomatic of a more generalized movement where a disruption to the assumed binaries between the developed and the developing, the eastern and the western is in evidence. The dis-location in this respect is threefold: India is emerging from its assigned 'Third World' location as a 'global player'; India is innovating in the arena of biotechnology; India is seen as a 'maverick' location devoid of high, exacting standards of ethical and moral governance as well as rigorous research and innovation. These three dis-locations are variously drawn on, together with empirical illustrations, throughout the book.

Chapter 2 attempts to locate the first dis-location by summarizing the global interest in India's so-called dizzying rise to sustained economic growth and the dramatic turn in its political fortune. This is briefly contrasted with the humbling poverty and a dismal record in basic curative care in the country. However, these marked differences are not so much a feature of India's failure to distribute and deliver fruits of economic growth evenly, but are rather a feature of her active embrace of the neo-liberal economic and political doctrine that, in part, is leading to the emergence of a neo-liberal moral economy. The chapter argues that when morality is introduced in the service of producing economically viable entities, it has a sanitizing effect on both production and consumption of entities (such as stem cell lines). It renders them ethically untainted, thus facilitating guilt-free consumption. To understand better the emerging moral economic context, the chapter attempts to unpack the second dis-location by describing the cultural context shaping post-colonial investments in science and technology research. Two seemingly distinct but connected themes are discussed. First, an examination of the images of science and technology and of 'biotech India' entertained by the political elite. Second, an analysis of the institution of a modernity that is ironically 'archaic', enabling a unique dialogue between an imagined past and the lived present. Here, one finds the centrally important role played by religion, myth, and philosophy, and their deployment to both promote and restrict embryonic stem cell generation and application. The chapter concludes by arguing that the seemingly simultaneous existence of the two antithetical 'Indias', one exotic and traditional, the other new and modern, is unhelpful. The reconfigured notions of tradition and modern are best understood as cultural responses to the process of rapid growth and change.

Chapter 3 is an encounter with the third dis-location that enables bio-crossings in and between spaces, such as nation–states, biology, law, philosophy and ethics. Central to the chapter is a (re)engagement with the notion of 'biosociality' as proposed by Paul Rabinow (1992). The chapter argues that the concept of biosociality, when viewed in the rapidly dis-locating context and its consequences in locales such as India, has unexamined limitations. When first proposed, the concept of biosociality was intended to conceptualize the emergence of 'associational communities around particular biological conditions' (Das and Addlakha, 2007: 128). Subsequent research has documented the existence of such communities (e.g. Rapp, 2000; cf. Gibbon and Novas, 2008). However, viewed from a global perspective, the concept continues to remain problematic. The chapter argues that biology is better understood as 'available' rather than 'social', especially when viewed in the context of bodily extractions and insertions achieved through a process of bio-crossing. In the Indian context, it is more fruitful to analyse identities, whether biological or social, more often assigned than assumed. Any identity politics in this respect is a struggle against an ascription (whether or not stigmatizing), rather than the assumption of a new identity that articulates resistance. The chapter further elaborates the very notion of biology as biologically-based biographies, and argues that an individual or institutional biography is inextricably inserted into the 'bio'. The chapter offers several ethnographic illustrations to make one central point: that bio-crossings achieved through extraction and insertion occur in zones that do not always fall within the normative arena of rule-governed and validated science. The struggles of the various protagonists, whether maverick scientists or individuals undertaking bio-crossings in search of health and healing, are almost always achieved in opposition to identities ascribed to them. As a result, they fail to acquire biosociality. The chapter concludes by reflecting on the very notion of structure and agency and proposes limitations that emerge in the dis-located, bio-crossed world of patients and practitioners.

Chapter 4 explores how embryos donors become enrolled in the moral economy of stem cell generation. It argues that the embryonic gifts and donations in India today are better understood if placed in their global contexts. These include: moral concerns surrounding embryos; concerns relating to purity and danger in the manufacture of biogenetic entities like stem cells; the governance ideals and ethical injunctions that promote regulated, governed and moral end products. The chapter describes how morality, together with other ethical and good manufacturing standards, is being globally outsourced in pursuits of research and profit.

Received ideals concerning ethical production are hastening India's participation in the global moral economy. They are also helping to create neo-liberal, consenting and choosing citizens. They in turn are imagined as gifting away their surplus embryos for research in an informed manner. The chapter proposes that, because of moral economic necessity, a notion of sacrifice may be more appropriate in understanding the accomplishment of the profitable circulation of the embryonic tissues. This analytic perspective is opposed to seeing them, in the words of Agamben (1998) as 'a killable bare life'. The chapter draws on the narratives of embryo donors, scientists and clinicians in research/clinical facilities in Delhi and Mumbai, as well as proposed ICMR guidelines, to describe how ethics, morality and consent are being accomplished as India accelerates its participation in the neo-liberal mode of production.

Chapter 5 takes the book into the liminal third space that can be derived by analysing local and global media interest in the proliferation of stem cell research in India. Focusing on the major tropes in the media narratives, the chapter describes how a twentieth-century view of the globe as essentially bipolar is resurrected in stories that oppose good and bad, ethical and maverick science. It examines the anomalous third space that goes beyond a simple distinction between local and global. This is a space policed by desire for and against liberal rules and governance. In other words, the network of actors/objects/discourses that make up the third space of ESC science in India (and perhaps elsewhere) is enveloped in a 'between time' that shapes the anti-structural nature of this spatial configuration. The chapter concludes by arguing that the criticism of the protagonists in the media stories promotes a certain image of global science, casting any notable departures from this imagination in pre-defined and culturally scripted in the roles of the 'maverick' and the 'renegade'. Ironically, in the Indian scientific terrain, there are historical precedents of such 'othering' tendencies and their media rendition. The chapter concludes by arguing that the *liminal third* compels a constant reordering by imploding opportunity and danger to rebirth the reproduced difference.

Chapter 6 addresses the ethical and legal governance of embryonic stem cells. The chapter discusses the growing international interest in maverick science and scientists and suggests that the well-documented trade in human organs may render India vulnerable to 'maverick stem cell science'. With the biotechnology boom in India as its backdrop, the chapter outlines stem cell regulation in global locales such as the United Kingdom and the United States of America in order to analyse the adoption of Draft Guidelines in India. The key area of informed consent is then discussed in an attempt, following Jasanoff (2005), to develop a deeper and more nuanced understanding of the ways in which normative discourses about biotechnology are embedded in wider culture and practice.

Chapter 7 recapitulates the substantive arguments of the book. It then invokes the concept of civic epistemologies to provide a more nuanced, comparative framework in which to situate the discussion of the local and the global. The book concludes by suggesting that India provides a useful basis for extending such concepts into cultures that, only superficially, appear to be far removed from Western democracies.

Local Cells, Global Science is not a definitive account of the global spread of embryonic stem cell research, or, for that matter, India's place in the emerging global circuits of scientific legislation, biological production, and distribution of tissues. It does provide a unique, culture-specific insight into the politics, economics and ethics surrounding the circulation of human tissues in the twenty-first century. India is one key example of the rapid proliferation of such biomedical phenomena around the globe.

2 Dis-locations

Local cultures of cells, global transactions in science

Introduction

This research explores the rapidly dis-locating landscape of biotechnological research around the globe. It is a landscape characterized by the spread of (bio)science and technology research to global sites that were, until now, not thought of as locations of 'biotechnological autoproduction' (Rabinow, 1996). Such dis-locations are rapidly problematizing long-established oppositions between the global North/South, First/Third Worlds, developed/developing economies and Western/Eastern cultures. Taking the rise of embryonic stem cell (ESC) research in India as its ethnographic focus, the book shows that what is to count as local and global is rapidly dis-locating. The twentieth-century development discourse that privileged the unidirectional flow of knowledge from the 'global' North/developed to the 'local' South/developing is now both an untenable orthodoxy and an unsustainable project. In the new century, the taken-for-granted, static locality of the 'South' and the inevitable global domination of the 'North' bleed into each other. In these emerging and dispersed biotechnological assemblages (Rabinow, 1999; Ong and Collier, 2005), ethical ideologies, governance protocols, free markets, venture capital, and geo-political cultures of scientific research, strategically frame global and local spatial configurations in unprecedented ways. These changes exemplify what Appadurai (1996) has described as the *technoscape*: that is, a fluid global configuration of technologies that moves at high speed across previously impervious boundaries (cf. Inhorn, 2003).

A biotechnological *technoscape* continually reassembles the global biological phenomena (Franklin, 2005). Discussing stem cell research in the UK, Sarah Franklin shows how, as a distinctive, emergent life form, stem cell production has become a global biological enterprise. She argues:

> Stem cells 'in culture' at the moment are being 'fed' by the production of norms, principles, values, and laws, as they are also being 'nurtured' by venture capital investments, media coverage, and public-sector funding. Certainly stem cells are being carefully tended by highly trained scientists, who are trying to teach them basic obedience lessons in state-of-the-art

laboratories from Singapore to Silicon Valley. They are being watched over carefully by presidents, prime ministers, and innumerable professional organisations concerned with their welfare, their rate of population growth, and their international travel arrangements. Few offspring have their provenance, ancestry, reproductive behavior, or genetic composition more carefully scrutinised by highly trained custodians.

(ibid.: 74)

The research on ESC materials in India is one salient node in this complex, shifting and (dis-)assembling, global space. India is a dis-location in multiple senses of the term. India has emerged as a global player alongside its neighbouring China. The emergence of Asia on the global scene challenges twentieth-century assumptions about First and Third Worlds. While India, in the past decade, is increasingly perceived by the world as an outsourcing haven for call centres and cheap information technology products, it is yet to be seen as a biotechnology hub. However, the silent but swift spread of bio-technology innovation and application in India is transforming the country into a preferred location for research and development. Innovation in biotech, in this respect, is no longer a 'western' prerogative. India is also a dis-location in that it is still vulnerable to being 'othered' as a global locale involved in unethical and maverick science. As we shall see later, this assumption is not only problematic but is, in the dis-located world of biotech invention and market-driven competition, an inevitable corollary of its status.

A dis-location is an ethical, moral, geo-political and economic digression from an imagined normative global order. It is a problematic spatiality, and matter out of place (Douglas, 2005). It is, by virtue of its liminality, a site of opportunity as well as of danger (van Gennep, 1960; Turner, 1967). It is a departure from a comforting status quo, and arrival into the uncertainty of an upheaval. It is nostalgia and the tantalizing future promise of the as yet unknown. To dis-locate is to aid the process of relocation and to contribute to the uninterrupted process of reassembling (local and global) spatiality. This book is an account of one such dis-located global site rapidly relocating itself in the new global order.

This chapter continues with a description of the rise of 'neo' as opposed to 'new' India. Using illustrations of India's participation in the global economy and the effects of this participation, the chapter examines the spread and consequences of neo-liberal political and economic calculations around the globe. The rapid rise of India's economic fortunes in the global sphere, and its dis-locating effect on the country's former 'Third World' standing, help to explain the power of the neo-liberal moral economy. The question asked is: What are the consequences when morality is incorporated as a key factor in economic and political calculations? It is argued that when a moral sensibility is invoked in the service of producing economically viable entities, it has a sanitizing effect on both the production and the consumption of entities (such as stem cell lines), by making them ethically untainted, and thus facilitating guilt-free consumption. It is within this broader frame that the culture of embryonic stem cell research in India is described. In so doing, this chapter offers a brief overview of post-colonial India's romance with biotechnology, the political and economic structures that have contained and sustained this engagement, and the fertile philosophical, religious and cultural thought processes that have enabled the engagement with biological materials and knowledge.

The rise of *neo*-India

The long-anticipated rise of India is now a reality. The International Monetary Fund, America's National Intelligence Council, the US Chambers of Commerce (which represents three million companies), and the UK's Chancellor of the Exchequer, have all warned of the 'growing strategic threat' emanating from India, which will be the next 'knowledge superpower'. Similarly, the UK news media, from national broadsheets to magazines such as the *New Scientist* and the *New Statesman*, regularly carry descriptions of India as the new superpower and the next knowledge economy. India is described as having had a meteoric rise from being a Third World country, to the brink of becoming the third largest economy in the world. The statistical evidence put forward in support of such claims is impressive. India consumes over 64 per cent of its own GDP, second only to the United States at 84 per cent and ahead of China at 50 per cent. It is estimated that, based on purchasing power parity (PPP) estimations, India will overtake Japan in a matter of months rather than years. This is because the Indian economy is growing at 7–8 per cent a year, a figure that is set to reach double digits by 2010.

In February 2006, the Bombay Stock Exchange's benchmark Sensex Index crossed the 10,000 threshold for the first time. As the Sensex moved from 9,000 to 10,000, foreign investors pumped $3 billion into the market, and the Sensex has since passed the 13,000 mark (at the last count, this figure stood at 20,000). The scope of India's rise can also be identified from the fact that, for the first time in its post-colonial existence, Indian investments in the UK have exceeded the reciprocal flow from its former colonial master (Bawden, 2006). Indian companies have invested $1 billion in new British developments, or greenfield projects, and this figure is set to rise. At the 2006 World Economic Forum meeting in Davos, India, launched the biggest charm offensive in the history of the Forum. Run at the cost of US$5 million, the Indian message was simple: 'India Everywhere'. The extravagant display of India's new-found confidence was everywhere, from billboards, through parties serving Indian food and drink, to Bollywood evenings. A contingent of forty-one senior executives and top cabinet ministers, including the finance minister, repeated one message: 'You can't ignore India any more.' One billboard encapsulated the message to the world – '15 years, six governments, five prime ministers, one direction' – signalling India's commitment to pursuing neo-liberal market reform.

This transformation is a far cry from the early, post-colonial pattern of growth, in which India's mixed economy model produced a steady annual rate of growth of 3.5 per cent, euphemized as the 'Hindu growth rate'. Since the liberalization of the Indian economy in 1991, India has capitalized on many fronts. The country has seen the spectacular progress of the information technology, biotechnology and pharmaceutical sectors, the opening up of the retail sector to foreign investment, the rise of home-grown multinational corporations and Fortune 500 blue chip companies, and a successful space research programme that promises to deliver the first Indian moon landing by 2010. Few will doubt that 'neo-India' has had a successful launch. However, neo-India is also very poor. In India, nearly 300 million people live on less than $1 a day (World Bank, 2006). Almost a third of the world's poor are Indian citizens. India has a one million-strong IT workforce that earns roughly 65 times more per head than its 400 million farmers (Luce, 2006). The education system in India has failed the almost 39 per cent of the country's population who can barely read or write, but at the same time its Indian Institutes of Technology and the Indian Institutes of Management have emerged as global centres of excellence in technical training. The health of the nation is equally worrying, with 47 per cent of under-fives underweight (UNICEF, 2006). Despite dramatic increases in innovation-led biotech products and generic drug manufacture, the basic health care needs of millions of Indians are either unmet or poorly addressed (Nichter and Nichter, 1996; Bharadwaj, 2005b). It is in this neo-India that biotechnologies of assisted conception and embryonic stem cells have begun to flourish (further described in Chapter 6). This flourishing of stem cell research, however, is, in part, symptomatic of an emerging neo-liberal global order that views ethical and moral calculation as key to political mobilization and economic articulation.

The neo-liberal moral economy

Aihwa Ong has recently conceptualized neo-liberalism as a new relationship between government and knowledge, through which governing activities are recast as non-political and non-ideological problems that require technical solutions (Ong, 2006: 3). She argues that in the neo-liberal model, market-driven considerations infiltrate the domain of politics and governance, and reconfigure the relationship between governing and governed, power and knowledge, sovereignty and territoriality. Conceptualizing neo-liberalism as exception, Ong suggests a space for an extraordinary departure in policy that can be deployed to include as well as exclude subjects and citizens. According to Ong, 'the politics of exception in an era of globalisation has disquieting ethicopolitical implications for those who are included as well as those who are excluded in shifting technologies of governing and of demarcation' (ibid.: 5).

In a globalized world, therefore, the neo-liberal market-led logic of exception has impacts on lives, rights and ethics in an ever accelerating manner.

This creates terrains of moral and ethical uncertainty that must be (re) articulated, explained and contained through further economic calculation, technological application, and a generalized sentiment of purportedly value-free governance intervention. The social and political predicament of 'subject–citizens' under the neo-liberal state of exception, on the other hand, is best appreciated in the words of Paul Farmer who contends that under the neo-liberal doctrine:

> [I]ndividuals in a society are viewed, if viewed at all, as autonomous, rational producers and consumers whose decisions are motivated primarily by economic or material concerns. But this ideology has little to say about the social and economic inequalities that distort real economies.
>
> (Farmer, 2003: 5)

The economic and scientific rationale underpinning the contemporary neo-liberal formation, therefore, compels an assessment of the moral and ethical corollaries of production and consumption practices, and the relations and forces of productions that sustain these activities. There is also the attendant problem of governing and regulating the markets in a rapidly globalizing and dislocating world, in which purportedly autonomous and rational producers and consumers interact.

The question of ethics and morality under neo-liberalism becomes particularly vexing when the *biopolitics*, the regulatory controls exerted over populations and individuals in order to harness and extract life forces (Ong, 2006), are considered. It is not surprising therefore, that developments in the field of biotechnology where, as Franklin shows, a discernible 'shift from kind to brand' is very much in evidence (Franklin, 2003b), underscore this shift with market-led calculations for ethical production and consumption. Dealing with the emergent dimension of innovation and market strategy in the business of stem cell manufacture, Franklin (2003a) has described ways in which concerns about public opinion are being built into new life forms. Franklin invites us to question 'what it means for social viability to become part of producing biological viability'. She locates public reactions to the cloning of Dolly the sheep, between the twin poles of public reaction to biotechnology that mingles hope and pride in the achievement with fear of its consequences (ibid.).

The unprecedented ethical complexity of the new biology under the neo-liberal formation also creates states of exception, where capital, profit, and accumulation, as well as production, consumption, and circulation, by (market-dictated) necessity become aligned to local/global ethical and moral registers. Adriana Petryna (2005) has recently argued that, in the context of an emerging industry of human-subject research, 'one can observe how deliberations over the ethics of research in crisis-ridden areas are set against – even eclipsed by – the market ethics of industry scientists and regulators'. Thus ethics as 'ethical variability', Petryna argues, becomes the industry norm and is even consciously deployed in pharmaceutical development (ibid.: 192).

The search for ethical and moral standards is, in turn, inextricably linked to the creation of standardized research, assembly and production protocols, routinized manufacturing practices, and the purchase and procurement of raw materials and expertise contained in standardized supply chains around the globe. This seemingly homogeneous process, in fact, exists in the moral, ethical and institutional state of exception. While research practices, manufacturing processes, labour markets, quality controls, and moral, ethical and governance protocols, are mobilized to create a standardized entity, whether an embryonic stem line or a pair of shoes, political formations and businesses are seldom able to achieve the 'ethical end products' without invoking states of exception and dispensation. This neo-liberal condition can be described as a *moral economy*. This demands that one asks, What kinds of regimes of exploitation, tropes and technologies of obfuscation, and processes of unproblematic, guilt-free consumption, are facilitated if the 'moral' in the economic calculation of new invention, experimentation and globalization is interpreted literally?

The notion of moral economy emerges in E.P. Thompson's (1971) seminal essay 'The moral economy of the English crowd in the eighteenth century'. Thompson's conception of moral economy was confined to confrontations in the market-place over access to necessities, profiteering and the marketing of food in times of dearth (Edelman, 2005). The concept was refined and elaborated by James Scott, who theorized peasant movements in South-East Asia along similar lines, and provided insights into the everyday resistance practices of the weak (Scott, 1976, 1985). Subsequent work using the theory has expanded on these original concerns, most notably developing an experiential theory of exploitation (Edelman, 2005). This argues that, 'under the prevailing neo-liberal dispensation, the nation–state has been largely superseded as the locus of market-rule-making by the WTO, the IMF, and the World Bank' (Scott, 2005). This view, however, overlooks the various states of exception that nation–states are able to create in order to not only resist, but also meld with, the governing sensibilities of like-minded nation–states in order to better serve their purposes, and to remain economically competitive among other global neo-liberal formations.

In order to understand neo-liberal states of exception better, the notion of moral economy could be gainfully read literally, as economic activity hedged with moralistic reasoning and conditioning. To make the economic terrain moral is to facilitate exchange, surplus, and accumulation that is seemingly untainted and guilt-free. In arguing that there is an emerging moral economy of embryonic stem cells, the book refers to the process of creation, isolation, extraction and exploitation of human embryos to generate and circulate economically viable and globally dispersed biogenetic stem lines. The fashioning of a global 'moral consensus' enables entities that are contested and ethically tainted in many cultural contexts (most notably the Euro-American countries) to profitably produce stem cell lines. For the moral economy to flourish, the state of exception must either remain systematically mis-recognized, or it must become a feature of the autonomous, rational, informed and free thinking ideological calculus of neo-liberal citizenship. The moral economy of a *technoscape*, therefore, not only facilitates guilt-free consumption, but also renders politically viable a dimension of human bio-commerce that, until now, has remained located at the intersections of diverse religious, moral and ethical interpretations, objections and contestation. What are the strategic and practical implications for Indian biotechnological ambitions when viewed in the context of the neo-liberal moral economy? A complex of cultural, social, political and economic developments links India to the neo-liberal moral economic circuits. These developments have been in the making for the better part of India's post-colonial history.

Cultures of embryonic stem cell research in India

There is ongoing worldwide scientific and media interest in India's burgeon-ing stem cell technology research programmes. The scientific advance in stem cell research in India has been described as the 'next big thing to hit India after the country's software revolution' (Thorold, 2001). With tight restric-tions imposed on funded research in the United States, and only marginally less restrictive regimes in the United Kingdom and elsewhere, India is increasingly being seen as 'filling the void in stem cell research', prompting speculation in segments of the Indian news media that this has effectively opened up a new 'pot of gold' for Indian science and business (cited in the *Washington Post*, 2001).

The main supply of stem cells in India comes from unused embryos created by the 'test-tube baby' clinics in the largely unregulated private sector. Under the Indian Council for Medical Research Guidelines, it is legal to use embryos up to 14 days old in medical research. A lack of formal regulation makes it easy for Indian scientists, unlike their Western peers, to forge ahead relatively unencumbered. In biotech research, there are moves afoot to reg-ulate these technologies by enshrining the Guidelines in legislation. This highlights how the recent emergence of embryonic stem cell research in India is a good example of cross-cultural differentials and the dispersion of research in biotechnologies. Following Appadurai (1996), it is useful to con-ceptualize the Indian case as an illustration of the *technoscape*, although it is important to note that technologies are never transferred into cultural voids. Local considerations, whether cultural, social, economic or political, shape and curtail the way in which western-generated technologies are both offered to, and received by, non-western subjects (Inhorn, 2003). Equally, (bio)tech-nologies embody complex cultural patterns that journey with them when they travel across cultural borders. This two-way cultural crossing drives a technology to *imbricate*. Diverse overlapping cultural forces shape how a technology is received, debated, used and promoted (Bharadwaj, 2005b).

International collaboration in developing new biotechnologies has seen exponential growth in recent years, marking a shift from local or national innovation systems to global innovation systems (Bartholomew, 1997). There is growing evidence to suggest that this move has bypassed firms in developing countries (Buctuanon, 2001). It is argued that developing, or transitional, economies lack not only the technological infrastructure, but also the scientific know-how necessary for biotechnology-related research (Zilinkas, 1995). The Indian experience has been different. In India, biotechnology is one of the fastest growing sectors. According to the Confederation of Indian Industry (CII), the Indian consumption of biotech products was US$1,789 million in 1999, and is expected to grow to US$4,270 million by 2010 (CII, 2003).[1] In addition, as will be discussed later in the book, the value of the medical biotech market has grown exponentially from the mid-1970s. According to a McKinsey study, the Indian pharmaceutical industry was poised to grow from the US$5 billion generic drug-based industry, into a US$25 billion innovation-led industry by 2010, with a market capitalization of almost US$150 billion (ibid.). However, even this dramatic increase is unlikely to have an impact upon the health care needs of India's large and growing rural population. Governments in post-colonial India have emphasized rural health care in successive Five-Year Plans, in the shape of infrastructural investment and programmes for primary health care (PHC). However, due to apathy and neglect, the infrastructure cover has made little difference, and rural PHC provision remains poor, with substandard facilities, inadequate supplies, poor management, a lack of monitoring, and the promotion of the personal interests of PHC staff (Banerji, 1974; Stark, 1985; Boerma, 1987; Nichter and Nichter, 1996). India's development in the arena of health biotechnology and stem cell production is, therefore, often seen as being at a crossroads (Frew *et al.*, 2007). The remit for biotech advances in India 'must not only address the significant health needs of its domestic population, but also position itself to take advantage of the often more profitable global marketplace' (ibid.: 403).

Recent commentators have suggested that, while in the arena of stem cell development India might challenge Western-centric notions of commercialization, a lasting constraint may lie in India's governance of basic stem cell science, clinical experimentation and drug trials (Salter *et al.*, 2007: 87). This view is not dissimilar from the emergent, neo-liberal imagination of a global order, where science, commercialization and profit become viable only so long as a universal ethical and moral calculus is inserted in the production and consumption process. Just what might this ethical and moral reasoning look like in the dis-located global landscape is difficult to speculate upon. If the rise of India and China, as well as their predicted growth trajectories, is taken as an important node in this ongoing re-location of the global sphere, then the locus and focus of ethical and moral reasoning may still be up for (re)negotiation, dilution and bargaining. India might truck, trade and barter on some of these crucial issues for the foreseeable future, in order to facilitate foreign direct investment, but this only exposes the contingent nature of governance and ethical thinking that fuels guilt-free consumption under the burgeoning neo-liberal moral economy.

The ongoing programme of drug trials in India is a good illustration of this tendency. According to a Confederation of Indian Industry study, Indian clinical trials in 2002 generated US$70 million in revenues. The study predicts that these revenues would grow to US$200 million by 2007, and to anywhere between US$500 million and US$1 billion by 2010. Big global pharma names, such as Novo Nordisk, Aventis, Novartis, GlaxoSmithKline, Eli Lilly and Pfizer, have begun clinical drug trials in various Indian cities (Rediff, 2004). For the first time, the Indian state is attempting to generate biovalue by *reinvesting* its surplus of viable citizens in the liberalized and booming economy. The fact that this (re)investment of surplus citizens is seen as morally unproblematic validates the process. The trials are reportedly conducted by fastidiously keeping to the twin (neo-liberal) legitimating mantras of 'informed consent' and 'ethical review'. This has a sanitizing effect on the process of clinical trials, bestowing official legitimacy through state-sponsored ethical guidelines and safeguards. In reality, however, there is growing evidence to suggest that the majority of participants remain unclear about the exact modality of administering drugs and the intricacies involved in consenting to the received information (BBC, 2006). This is further discussed, in the context of stem cells and IVF clinics, in Chapters 4 and 6.

Sunder Rajan has recently shown that clinical trials in India are not conducted in the absence of ethical framing or informed consent, but rather in a space of 'hyperethicality', where trials are tightly regulated under some of the strictest clinical trial laws in the world (Sunder Rajan, 2007: 179). Therefore, the most prominent and internationally recognized clinical research organizations (CROs) in India are not so much illustrations of dubious and unethical experimentation on human subjects, but, rather ironically, are shining examples of gold standard research. In these ethical spaces, healthy experimental subjects are constituted (for Phase I trials) as 'informed' because they are viewed as being among the most compliant, with better retention rates as compared to experimental subjects in the USA. In addition the cultural attribute ascribed to Indian trial volunteers (and others) is that 'people trust doctors here' thus producing a 'merely risked experimental subject' outside the circuits of 'pastoral care and therapeutic consumption' (ibid.: 177–9). Thus, for Sunder Rajan, the violence implicit in clinical trials in India is not in an imaginary realm of absent ethical framing or its violation, but it is rather structurally embedded in the ethically informed and legislated system itself. Here, informed neo-liberal subjectivity gets constituted as merely risked by the provision of information, and perhaps also through the very specific nature of a 'culture of trust' that enables 'rational subjects' to 'trust' the information, as opposed to questioning the knowledge base upon which both the trial and the experimental subjects are predicated. The discussion in Chapter 6 takes this starting point to offer an analysis of what such a culture of trust actually entails.

The development of stem cell research and application in India is part of a larger policy that has been pursued by the Indian state since 1985. The Indian government invested in the creation of a dedicated central Department of Biotechnology (DBT) as the nodal agency for policy, promotion of R&D, international cooperation, and manufacturing activities. The Indian government has invested nearly US$500 million in this venture. The Department of Biotechnology has launched two major programmes of stem cell research, aimed at treating blindness and certain brain disorders, and a significant amount of funding has been set aside for the Indian government's Tenth Five-Year Plan. With the establishment of a private Indian biotech company, Reliance Life Sciences, in Mumbai (formerly Bombay), public and private sector research institutions are forging ahead with transnational collaborations. In early August 2001, the United States National Institute of Health announced that Reliance Life Sciences, along with the state-owned, Bangalore-based National Centre for Biological Sciences, would be among the ten institutions world-wide that would receive federal funding for stem cell research. Both centres have met the US administration's criteria for the derivation of human embryonic stem cell lines. On 9 August 2001, US President George Bush allowed federal funding of embryonic stem cells research to go forward on cells that were already in existence (discussed in Chapters 4 and 6). Around the world, only 64 cell lines could meet this criterion and, of these, ten existed in India. Three lines were held by the state-owned National Centre for Biological Sciences, Bangalore, and a further seven were under development at the private research laboratory of Reliance Life Sciences in Mumbai (see Chapter 3).

This private–public sector symbiosis, however, is not specific to the development of biotechnology in India. As argued elsewhere, despite its avowed commitment to providing health care and planning, the post-independence Indian state did not curtail private health provision and, under the shelter of a mixed economy model, the private and public sectors have managed to establish a protracted symbiotic coexistence (Bharadwaj, 2001). This public–private symbiosis, however, could not flourish in the biotech arena where the public sector led the way due to the vast resource-base of the state. A good early example of this is the setting up of BIOCON, India's first biotech company, in 1978, followed by the Centre for Cellular and Molecular Biology, for DNA- and r-DNA-based research, in 1981. Such state investments in research institutes, laboratories and public hospitals could not be matched by the private sector until the 1990s 'liberalization era'. Private capital investments produced not only the dawn of corporate hospitals, described by Relman (1987) as the 'medical industrial complex', but also the launch of private biotech research companies such as Reliance Life Sciences.

In India today, there are two prominent forces connected to the rise of biotechnology research and development. First, a political investment in the very idea of what a 'biotech revolution' might accomplish for the country, especially since a vision of 'biotech India' chimes with the nationalistic imagining of a developing and resurgent India. Second, the cultural and religious context in India places the development of so-called red biotech in a uniquely anomalous and contested space. Religion, myth, and oral histories are being resurrected to promote or oppose the idea of embryonic generation and manipulation in the service of science, development and profit. In what follows, this chapter briefly analyses these two seemingly opposing but intimately connected domains of politics and religion, and demonstrates how, together, these supposedly opposing terrains enable embryonic stem cell production, and science more generally, in the local moral spaces in India.

IT to BT: imagining biotech India

The political and economic developments described above resonate with the political elite in India. Not surprisingly, the most commonly cited observation on these developments is that of the former Indian Prime Minister, Atal Bihari Vajpayee. Vajpayee described India's success in the information technology (IT) sector as 'India today', and BT, i.e. biotechnology, as '*Bharat* Tomorrow' (Bharat being the Sanskrit/Hindi name for India). The shift from India to *Bharat* and from English to Sanskrit/Hindi is especially interesting. It is possible to conceptualize this ideological and semantic shift on the plane of the everyday and ordinary, where an acronym is employed to imagine India as a resurgent global player, and on the plane of the political, which was, at the time, dominated by the Hindu nationalist-led coalition government headed by Prime Minister Atal Bihari Vajpayee of the *Bhartiya Janta Party* (BJP), and its affiliated ideologues *Rashtriya Sevak Sangh* (RSS), whose stated vision is to see India as a global Hindu superpower. The general elections for the Indian parliament in April/May 2004 saw the defeat of the BJP-led coalition government. The verdict was widely analysed as a rejection of the BJP's elitist polices by the rural poor of India, which sent a coalition government headed by the Congress Party to power. The new government initiated new policies, but continued the biotechnology-based agenda of its political rivals. By November 2005, the government had explicitly stated its intention to develop both short- and long-term strategies for research and development on stem cells (*The Financial Express*, 2005). It is often argued that the BJP-led Indian government was responsible for much of the 'economic liberalization' of the country, which was kick-started by the Congress Party government in 1991, by dismantling the socialist policies of previous governments. Some commentators argue that, while this may be seen as a move to the right in the political spectrum, the policies of the BJP emulate the more open, capitalist societies of the West, but retain a *dharmic* background (Frawley, 2001). (In Hinduism, dharmic or dharma refers to the moral and ethical aims of human life, i.e. righteousness. In Buddhism, dharma refers to the content of the teachings and wisdom of the Buddha.)

In the context of biotechnology, this 'post-1990s ideology of economic liberalization', as routinely articulated in Indian elite and policy circles, has come to both visualize and idealize India, transforming it, in the words of Sunder Rajan, 'as India Inc.' (Sunder Rajan, 2002: 285). What can explain India's romance with the scientific high culture of biotechnology, given that the primary curative health care needs of millions are as yet unmet? Viewed in the larger context of nation-building and strategic positioning in the community of nations – spurred on by the liberalization rhetoric of the 1990s – the huge state investment and interest in biotech do not seem to be overly misplaced. The archaeology of these developments yields historic echoes that Gyan Prakash has documented in his portrayal of the intimate relationship between science and colonialism, and the rise of a peculiarly Indian modernity. In the context of colonial science, a prime instrument of the empire, Prakash shows how the rise of science came to instil in many of the colonized elite a need to seek 'parallels and precedents for scientific thought in India's own intellectual history, creating a hybrid form of knowledge that combined western ideas with local cultural and religious understanding' (Prakash, 1999: 7). According to Prakash:

> With great ideological imagination and dexterity the nationalists argued that Indian modernity must be irreducibly different, that the modern configuration of its territory must reflect India's unique and universal scientific and technological heritage. Thus, the Indian nation state that came into being in 1947 was deeply connected to science's work as a metaphor, to its functioning beyond the boundaries of the laboratory as a grammar of modern power.
>
> (ibid.: 7)

Two important developments emerge in post-colonial India as a consequence of this historic contact with 'Western science'. First, for almost forty years, conspicuous technology emerged as the official goal of science in India, an era often characterized as the Congress Party-led Nehru–Gandhi dynasty (Nandy, 1996). Over a period of time, both state and private media propaganda, and the values propagated by the westernized education system, instilled in the Indian middle classes a view of science as 'spectacular technology' (ibid.: 7). Second, while the historic, local (cultural) response to the global (science) was established in the context of a struggle for independence from British rule, it nevertheless laid the foundations for a brand of religious nationalism that sought to resurrect a 'glorious aura of a grand civilization forgotten by its own people' (Subramaniam, 2002). Thus, in post-colonial India, the religious nationalists (led by the BJP government) have succeeded in bringing together a modern vision predicated on science (the most prominent icon of which is India's nuclear programme), with an archaic vision of a glorious Hindu past, to create an 'archaic modernity' (Subramaniam, 2000). Subramaniam argues that the religious nationalists in contemporary

India have – by strategically using the rhetoric of science and Hinduism, modernity and orthodoxy, western and eastern thought – built a powerful, but potentially dangerous, vision of a Hindu nation. This archaic modernity does not see Hindu India as one basking in the timeless glory of its rich and imagined *Vedic* past but, rather one that has embraced capitalism, western science and (bio)technology. (The term 'Vedic' relates to the sacred texts called Vedas or, sometimes, to the ancient Sanskrit in which these texts were written. The Vedic literature is thought to have emerged in approximately 1500 BCE.)

In this respect, the desire for excellence in cutting-edge science (for example, stem cells), is driven by an ambition similar to the nationalistic fervour that frames state rhetoric around space exploration and the race to put an Indian on the moon (Harding, 2003). For that matter, it parallels the ambition to suffuse the world with cheap technical know-how and information technology. It is through these expressions of desire for scientific and economic domination that the Indian state is seeking to bring about a fundamental shift in its global political standing, desires that are most graphically literalized in its long-standing ambition to acquire a permanent seat on the United Nations Security Council, and more recently by acquiring its first overseas military base, in the central Asian state of Tajikistan (Walsh, 2006).

The archaic and the modern: religion, philosophy and biotechnology

Politically charged enunciatory tropes in India often draw on a notion of Indian civil society as being structured by deep-rooted 'cultural' ideas – seen as conducive to the conduct of science – that frame India as a fertile terrain for biotech production, promotion and consumption. However, as described above, in contemporary India these ideas have become an integral part of the nationalist project of reinventing India by selectively drawing on traditions in the service of an archaic modernity. While there is a notable absence of public debate or outcry with regard to embryonic stem cells, both dissenting and consenting voices have emerged from within the Hindu domain, despite the fact that stem cell biotech appeals to the religious nationalism and nation-building strategies of the ruling elite.

In August 2001, the first dramatic salvo was fired at the proliferation of embryonic stem cell research in India by one of the five leading titled Hindu leaders in India, *Shankaracharya* Jayendra Saraswat. In a written statement, the spiritual leader denounced interventions such as abortion, artificial insemination and test tube babies as 'sinful acts' and deemed them to be totally unacceptable. On the issue of stem cells the statement simply states 'from the stage of embryonic stem cell, life starts' (LifeSiteNews.com, 2001). Ironically, this objection to medical interventions in human reproduction and to embryonic manipulation is rooted in the same 3,000-year-old Vedic tradition that religious nationalists have consistently drawn on in order to shape the 'archaic' Hindu modernity. In the Vedic conceptual domain, five cardinal sins, *Pancha Maha Pathakas*, are recognized (Azariah, 1997): *Stree Hatya* (killing of women); *Go Hatya* (killing of cows); *Bhruna Hatya* (killing of embryos); *Brhama Hatya* (killing of Brahmin); *Shishu Hatya* (killing of babies). In Vedic society these transgressions were seen as threatening the very existence of a viable community.

Following Lévi-Strauss (1969) (and his feminist critics), it can be argued that exchange of women and their reproductive potential was central to social reproduction itself, and that any threat to this fertility, including its visible symbols, living children and viable pregnancies, were anathema to a society with high incidences of pregnancy loss and maternal and infant mortality. This is discernible from the elaborate records of Vedic rituals and religious ceremonies intended to ward off the dangers that surround pregnancy and childbirth (Bhattacharji, 1990). The cultural mandate to reproduce, therefore, construed any neglect in this regard as tantamount to killing embryos. This is evident from ancient legal codes such as *Manu Smriti*, the *Laws of Gautama* and the *Institutes of Vishnu*. These codes have influenced and governed civil society, to a great extent, on the Indian subcontinent; they are thought to date from between 4000 BCE to the mid-sixth century BCE: 'Reprehensible is the father who gives not (his daughter in marriage) at the proper time; reprehensible is the husband who approaches not (his wife in due season)' (*Manu*: IX, 3, 4, 328, Müller, 1894 trans.)[2]

The rules for the father – the bride giver – are, however, more strictly encoded in various ancient legal texts. The laws of *Gautama* for instance state: 'A girl should be given in marriage before (she attains the age of) puberty. He who neglects it, commits sin' (*Gautama*: XVIII, 21, 22, 269, Müller, 1894 trans.).

The code of *Vasishtha* goes a step further and states:

> As often as the courses of a maiden, who is filled with desire, and demanded in marriage by men of equal caste, recur, so often her father and her mother are guilty of (the crime of) slaying an embryo; that is a rule of the sacred law.
>
> (*Vasishtha*: XVII, 71, 92, Müller, 1894 trans.)

It is not difficult to see why the killing or harming of *Brahmins*, who not only authored these modes of thought and codes of conduct but enforced them too, was also seen as a cardinal sin. In this agrarian and pastoral economy, any threat to society's ability to perpetuate itself was understandably perceived as transgressing socio-cosmic order. This order was most visible in the dominant currency of the time – head of cattle – hence the importance of cows in India. The references to the killing of embryos and of potential life itself in these ancient records are, therefore, intimately rooted in a culture's struggle for social, political and economic viability. It is fascinating to find this elaborate historical record to be attached to the debates surrounding contemporary stem cell research. This is especially so given that it is just one fragment of a vast tapestry of customary law, and religious, moral and ethical philosophy, that actors in contemporary India have chosen to draw on, in opposing or supporting biotech research on embryonic material.

Another example commonly encountered is from the realm of *Upanishad*[3] philosophy (dated between 1000 BCE and 600 BCE). M.G. Prasad, head of a New Jersey-based Hindu temple, in responding to an ethical and religious debate in the United States, put forward the 'Hindu perspective', arguing:

> Dharma says that unless a complete understanding and realisation of answers for fundamental questions – What is life? When does life begin and end? – is obtained, any action and development in controversial research such as embryonic stem cell research would result from an incomplete knowledge of life. In dealing with discussions on embryonic stem cell research, a similar question comes up between a seed and tree. A seed contains a tree invisible to the naked eye. When does a tree become a tree?
>
> (Prasad, 2001)

This reference to the tree and the seed is an oblique reference to the *Chandogya Upanishad* (one of the 108 *Upanishad* texts), which dialogically muses over the complexity inherent in the universe:

> 'Bring hither a fruit of that nyagrodha tree'. 'Here it is, Venerable Sir'. 'Break it'. 'It is broken, Venerable Sir'. 'What do you see there?' 'These extremely fine seeds, Venerable Sir'. 'Of these, please break one.' 'It is broken, Venerable Sir'. 'What do you see there?' 'Nothing at all, Venerable Sir'. Then he said to him, 'My dear, that subtle essence which you do not perceive, verily, my dear, from that very essence this great *nyagrodha* tree exists. Believe me my dear.'
> (*Chandogya Upanishad*: VI. 13.1; XII, 462. Radhakrishnan, 1978 trans.)

In this passage, the *Upanishad* is seeking to explain how the apparent immensity of the entire universe is conceived to arise from the inner self just as a great tree rises from the unseen essence within a tiny seed (Wood, 1996a: 27). What is therefore encapsulated within this passage is an instruction to a pupil, enjoining her to see how, compared to the universe, individual personalities are small and insignificant (like a tiny seed of a large tree), and yet like trees growing from within the unknown depths of tiny seeds, our knowledge of the world arises from within our body and mind (Wood, 1996a, 1996b). This argument, however, is misappropriated to suggest, at least analogically, the potentiality of embryonic life. While the passage can become the philosophical basis for engaging with the question of scientific knowledge itself, it cannot be reconceived to suggest that, since from within embryos potential humans may emerge, there is something sacred about such potential. The problem posed by the *Upanishad* is not in the realm of 'when does a tree/life begin?' but rather a philosophical musing on the nature of knowledge and cognition itself.

In contrast, there are perspectives that see in the Hindu tradition considerable potential to countenance embryonic stem cells research. The responses are varied; one perspective, that draws on the sacred Hindu text, the *Bhagavad Gita* (thought to date from 550 BCE), the Hindu God Krishna's sermon in particular, contends that embryos can only be seen as interim stages of physical manifestations of the soul. The idea that 'Soul (*Atman*) is unborn, eternal and ever-existing and is not slain when the body is slain', becomes the basis for claiming that, since destruction or annihilation of the physical form does not destroy the soul, experiments on embryos should therefore not be abhorrent to Hindus (Kumar, 2003). Another view, drawing on Hindu mythology and the story of sage *Dadhichi*, describes that when the sage's bones were sought by the gods to eliminate a demon, the sage sacrificed his body for the greater good, and such an act was not seen or condemned as suicide. Thus, the destruction of embryos in stem cell research could similarly be considered as an extraordinary, unavoidable act, and for the greater good (Tyagananda, 2002). The most extraordinary justification for stem cell research, however, came from a scientist, Dr B.G. Matapurkar, who conducted research using adult stem cells in Delhi's Maulana Azad Medical College. He argues that *Adi Parva* (first chapter) of *Mahabharata* (one of the great religious Hindu epics) gives a clear indication that the one of the two main protagonists, *Kauravas,* was born from stem cells. Not surprisingly, an article appearing in the English-language magazine *The Week* carried the following banner headline on 16 September 2001, 'Stem Cells: A Lost Science of India?'

Claims such as these provide an important insight into the way in which India is being constantly reinvented by the media, religious nationalists, political and scientific elites, as both ancient and modern. As Subramaniam argues:

> When religious nationalists [to which we can now add other interested parties like the media and scientists] invoke the Vedas or other ancient scriptures in the name of Hindu pride; their vision does not supplant western science, but instead it melds with western science.
>
> (2002: 6)

The preceding discussion also provides an insight into how the very notion of ancient Vedic culture can be resurrected by different continuants and interested parties in the service of narrow political ends. In using such local cultural metaphors and Hindu ethical notions, harking back to *Vedic* and *Upanishadic* philosophy, to either criticize or promote global biotech in contemporary India, a number of agents, ranging from the political elites, through religious leaders, to scientists, have entered the fray. It is premature to say if, when and how these voices will align themselves across religious, party political and scientific terrains, and whether these moves will culminate in a systematic framing of a definitive 'Hindu bioethic'. It is in this developing context, of socio-cultural reasoning, hype and nationalistic imaginings of techno-scientific advancements, that a new generation of market-led research in genomics, proteomics, transgenics, and stem cells has emerged.

Conclusion

This chapter has analysed various ongoing local and global developments that frame the rise of embryonic stem cell research in India, a nation in which these developments are being assimilated in multiple terrains. These range from the political visions of the religious nationalists, through the post-colonial political economy of biotechnology, to the ongoing attempts to institute governance protocols and the ethical framing of biotechnology within the Hindu religious domain. Debates surrounding the development and use of embryonic stem cells in the West have centred on the moral status and personhood of the human embryo. The Indian case is far more complex, where both religious nationalist visions of modern, democratic, technocratic, capitalist Hindu India, have come to co-exist with the oppositional voices that frame their resistance to the biotechnology of stem cell from within the same Hindu conceptual domain. However, an imagination of India as modern yet traditional is not a mere localizable process.

The economic rise of India has attracted similar global assessments, which frame the country as both ancient and yet modern. Two important illustrations of this global trend can be gleaned from the media and from a more 'independent' policy-driven analysis. A recent cover story in *TIME* magazine is a fine illustration of the tendency. Dedicated to reviewing 60 years of Indian independence from British rule, the story declares 'India charges ahead' (*TIME*, 2007). In the concluding article, historian and author William Dalrymple describes India's rise as 'business as usual', arguing that, far from being a miraculous novelty, India's rise is, in fact, a return to traditional global trade patterns (Dalrymple, 2007: 56). In the traditional pattern of the medieval and ancient world, argues Dalrymple, 'gold drained from West to the East in payment for silks and spices and all manner of luxuries undreamed of in the relatively primitive capitals of Europe' (ibid.: 56). Dalrymple argues that in 1600, when the East India Company was founded, Britain was generating 1.8 per cent of world GDP, while India was producing 22.5 per cent. 'By 1870, at the peak of the Raj, Britain was generating 9.1 per cent, while India had been reduced for the first time to the epitome of a Third World nation, a symbol across the globe of famine, poverty and deprivation' (ibid.: 56). To conceptualize the economic rise of India as 'business as usual' is also, perhaps, one potent way of imagining the intrinsic and indigenous nature of cultural and material production as well as circulation around the globe. If, as Dalrymple argues, India is returning to a historical norm, then there is something deterministic and predictable about its economic fate. The religious nationalists hark back to a glorious and ancient past, one that surpasses the present, in order to produce an imagination of the future that, in turn, might one day surpass the imagined past. The international media and scholars alike have begun to view the timeless and yet modern India in the same light. This is perhaps one of the many dis-locating consequences of a disassembling global order.

The policy analysis emerging from independent 'think tanks', such as the UK-based DEMOS, offers further evidence of the deeply rooted imagination of 'an India' riddled with contradictions. The DEMOS report describes the paradox of Indian science as one locatable between the opposing worlds of 'BMW science' and 'bullock cart science'. Described as 'the uneven innovator' (Bound, 2007), India's possible future scenario oscillates between a future of global science leadership and techno-nationalism, and a future of 'equity before excellence' (i.e. low-tech ingenious products for the Indian market with bottom-of-the-pyramid focus) and 'offshore science service'. The main thrust of the argument, however, remains an overtly essentialist one that sees India as a collection of opposing and seemingly irreconcilable differences. Nevertheless, the BMWs and the bullock carts travel on the same pothole-riddled roads, and jostle for space with a multitude of other competing 'lifeworlds' that make incomprehensible an imagination of global leadership in science and technology innovation. The dualism of the BMW and bullock cart conceals ideologically entrenched assumptions about the West and the East, North and South, developed and the developing, and, more disturbingly, rationality and religion. A quick checklist of BMW science reveals potent signifiers of imagined modernity in India. These predictably include Bangalore, democracy, rule of law, 45 million graduates, a modern affluent growing middle class, conformity to WTO and TRIPS, science as the answer to the development question, innovation increasingly important as a source of market differentiation as well as profit, and an English-speaking, western-educated scientific elite. This picture of modern and innovating India is held in opposition to Bihar (the so-called failed state of India), corruption, 500 million people dependent on agriculture for survival, ancient society, traditional scientific knowledge requiring new concepts of knowledge ownership, the importance of religion, astrology and superstition, innovation, ingenuity and adaptiveness as a daily necessity for survival, and around twenty officially recognized languages with numerous dialects (ibid.). The verdict of this report is that this divide has possible benefits for Indian science and innovation, so long as there is no expectation to see development proceed along Euro-American lines, instead remaining open to a distinctive mix of culture, past and present, tradition and modernity.

The main oversight in this argument is that it does not adequately dwell on the rapidly reassembling global sphere in the post-Cold War era. The bullock cart and BMW contradiction was sustainable in a bipolar world, in which India consciously championed the non-aligned movement and located both her politics and economics between the capitalist and socialist ideologies.

Though seen as lying within the Soviet sphere of influence, India remained, for over 40 years, firmly committed to the mixed-economy model that favoured an eclectic combination of the private and the public sector part-nership. This partnership was rapidly dis-located with the collapse of the Soviet Union and the ushering in of forces of 'liberalization'. The bullock carts did not disappear overnight, and BMWs did not take over Indian roads, but the two did not begin to co-exist to re-emerge as a new paradigm. What did change, and continues to change, is India's embrace of the neo-liberal doctrine as one of the many inventive ways to embrace globalization, transnationalism, and internationalism. A process that has set in motion the creation of markets, both local and global, consumers, as citizens and sub-jects, and ideologies, of choice, consent and rights. Under the neo-liberal state, economics, politics, governance, citizens, rights and duties can be forged into a multitude of differing permutations, invoking states of excep-tion – to religion, science, governance, rights, and so on. This is a process of nation-building and of melding into the global sphere by inserting or removing ethics and morality into/from ideologies, products, material and cultural practices, governance and legislative modalities. The imagination of a timeless India, standing still in space and time and continuing to co-exist alongside a modern, resurgent Indian nation, is unhelpful and misleading. Like both the secular and religious nationalist political elites in India, local and global media and political and policy analysts are all too readily seduced by the notion of 'two Indias'. The co-existence of the see-mingly modern and the traditional, of science and religion, among other things, are curiously new ways of making sense of cultural change, a process that is not necessarily antithetical. One does not contradict or negate the other but offers creative possibilities of enabling the process of change itself (cf. Bharadwaj, 2006a).

India is changing, and the force for change is located in her newly recon-figured institutions, until now deemed modern and/or traditional. The force of change is at its most evident and palpable in the dis-locating global landscape that is an ongoing geo-political, as well as politico-economic, reordering of the global order. The rise of embryonic stem cell research is one node in the rapid ascent of India in this dis-located world. If by *tech-noscape* we understand the swift move of information, knowledge, expertise and people across previously impermeable borders, then this process is not only reordering global spatiality and problematizing what is to count as local and global, but it is also enabling new circuits of economic and political calculation. That is to say, this movement at great speed across previously impermeable boundaries is enabled by an ethical and moral sentiment. This facilitates movement by inserting the moral into the economic and political calculus of nation–states, and into their will and ability to govern. In the final analysis it enables the *technoscape* itself to function at a level that requires some semblance of moral validation. This process, described as the moral economy, is rapidly suffusing global markets, shaping consumption practices, having an impact on production processes and procurement pro-cedures, and framing governance protocols and inter-state relations. The chapters that follow further explicate and ethnographically illustrate con-cepts such as local, global, dis-location, and moral economy. It remains to be seen how the political, religious and economic rationale of the Indian state will shape the rapid development of both India and the embryonic biotechnology of stem cells. If, however, the ongoing scientific developments and the state support of an imagination of 'biotech India' are viewed as two powerful future indicators, then embryonic stem cells may well become a significant sector of the Indian biotech landscape.

3 Biosociality to bio-crossings

Encounters with embryonic stem cells in India

Introduction

> In the future, the new genetics will cease to be a biological metaphor for modern society and will become instead a circulation network of identity terms and restriction loci, around which and through which a truly new type of autoproduction will emerge, which I call biosociality.
>
> (Rabinow, 1996: 99)

The social science imagination has variously conceptualized the constructed nature of the biological, organic, natural and their shifting meanings. Concepts such as *hybrids* (Latour, 1993), *juxtapositions* (Strathern, 1992) and *cyborgs* (Haraway, 1990; 1991) have, alongside the notion of *biosociality* (Rabinow, 1992), been used to problematize the *bioscapes* of late- or post-modern societies. These analyses have all, in various ways, made problematic the relationships between nature and culture. The categories that we have taken for granted as 'natural' have been increasingly disrupted. Biological nature is increasingly subject to human, cultural intervention. Human activity no longer investigates biological phenomena; it makes and re-makes them. At the same time, aspects of individual and collective identity may be formulated in terms of biological characteristics and attributes. The latter may include shared medical risks, or shared narratives of common genetic origins.

The insights of authors writing on the mutual influences of natural and cultural domains have enriched the social sciences conceptually and methodologically. Yet they privilege the existence of one among many other possible cultural biographies of human biology. They can, for instance, inadvertently privilege culturally-specific notions of biological citizenship or political discourse surrounding biomedical innovation. The very act of demolishing the hegemonic formulations of nature/culture oppositions in the Euro-American worldview has unwittingly led to the 'anthropologization' of an equally dominant model of the biological and the social that continually bleed into each other. Such conceptual tropes present hitherto unexamined complexities when globalized.

It is, perhaps, fair to say that the notion of biosociality was never intended to be a 'master concept' seeking to unpack conundrums and complexities associated with the human biological form and its social context around the globe. It is not simply the case that the notion of biosociality is inadequate and that we must now expand the formulation to include the 'excluded others'. Rather, what is important is the predictive voice in which the concept was first enunciated almost fifteen years ago. Many of the predictions have stood the test of time (Rapp, 2000) and some others await fruition. Two of the most important pronouncements about the future, encapsulated in the idea of biosociality, are well known:

1. In the future ... [the] likely formation of new groups and individual identities and practices arising out of [these] new truths ...
2. ... nature will be made known and remade through technique and will finally become artificial, just as culture becomes natural.

(Rabinow, 1996: 99–102)

It is difficult to conjure a temporal dimension; to engage with the future is to imagine it, just as to engage with the past is to reinvent it. In each case, the processes at work are those of prediction and remembrance. However, if time can be remembered and predicted, then clearly the present can only be actualized in experience. The futures predicted when the concept was first proposed have the potential to be actualized and experienced in contexts that are not universally available. Therefore, for the concept of biosociality to be viable, its cultural and temporal contours must be outlined. Rabinow takes the American Human Genome Initiative (AHGI) as one 'logical place to begin' an examination of changes to life in the context of new knowledge and power. The contours of his project are precisely drawn and defined. In this respect, his ethnographic question is clear. He asks: 'How will our social and ethical practices change as this project [the AHGI] advances?' The very idea of 'our social and ethical practices' is posited at the notable expense of excluding those who seemingly do not fit the Euro-American category. This observation is not so much a critique but rather an attempt further to emphasize that the idea of biosociality is limited in the global context. Perhaps this is because, as a concept, it was never intended to travel very far in the first place.

In global spaces such as India, biotechnologies are available and thriving, as opposed to locales both within India and elsewhere, where crippling poverty asphyxiates both the 'bio' and any semblance of 'sociality'. Such ethnographic 'truths' cannot be wished away if anthropology and other social sciences are to continue to understand, explain and criticize the 'difference' permeating the world. The conceptual and anthropological challenge, therefore, is to remain in an anticipatory space. A space that does not seek to preempt 'difference' through intuitively second-guessing its emergence, but rather is alert to its possible existence. In this respect, this chapter can be read on two levels: as a problematization of the concept of biosociality, and as a tentative step towards further elaborating the concept through critical thinking.

The question remains, 'How do people seeking biomedical interventions exist on terms far out of their control, certainly not of their choosing and seldom of their own making, in an emerging neo-liberal formation such as India?' The same question can be posed elsewhere. However, the question, and the terrain it instantiates, seem appropriate for an analysis that seeks to bring into focus the biosociality of individuals and potential groupings who find the lived experience of their bodies irreversibly scarred by neo-liberal statecraft and political economy. And yet, the complexities and compulsions individuals and collectivities face in emerging neo-liberal formations such as India (Bharadwaj, 2006c) seldom produce powerful opportunities to harness or gain anything remotely profitable from being biosocially active. The constraints are multiple, from the unavailability of opportunity, and of resources, the inability to organize around a medical condition, syndrome or mutation, to social isolation, stigma and de-legitimation.

This chapter, therefore, examines the concept of biosociality in its present shape and form, and proposes limitations that emerge when it is applied to a rapidly transforming India. In so doing, the bio is (re)conceptualized as *available*, as opposed to *social*, by focusing on the processes of extraction and insertion that not only generate 'biovalue', but also facilitate what will be described as *bio-crossings* (Bharadwaj, 2008).

Bio-crossing, extraction, insertion

While a lot of ink has been spilt discussing the neo-liberal agenda around the globe, Paul Farmer's (2003) formulation, as outlined in Chapter 2, is helpful in contextualizing neo-India. The distorted face of neo-liberal economies is further exemplified if we add to the picture the parallel bio-economies and, following Waldby (2002), the promissory biovalue they hope to generate:

> Biovalue refers to the yield of vitality produced by the biotechnological reformulation of living processes. Biotechnology tries to gain traction in living processes. To induce them to increase or change their productivity along specified lines, intensify their self-reproducing and self-maintaining capacities. This intensification or leveraging of living process typically takes place not at the level of the body as a macro-anatomical system but at the level of the cellular or molecular fragment, the mRNA, the bacterium, the oocyte, the stem cell.
>
> (Waldby, 2002: 310)

Such a way of capitalizing on life, according to Walby, produces a margin of biovalue, a surplus of fragmentary vitality. There are, according to her, two incentives in the production of biovalue. First, there is *the public incentive* or the biosociality of advocates hoping for the creation of a *use value*. A good example is the promissory future emerging from stem cell therapies. Second, there is an incentive to generate *exchange value* with biological commodities that can be bought and sold in a purportedly free market. However, whereas in capitalism capital is accumulated, in the biomedical enterprise, capital is promissory (Thompson, 2005), raised for speculative ventures on the strength of promised future returns (Franklin, 2003b).

Waldby's thesis identifies the generation of biovalue through the intensification of living processes at the molecular level, or levels of similar scale that require a microscopic gaze, as opposed to macro-anatomical systems. However, the very act of generating biovalue makes the macro-anatomical system valuable in the market-driven, neo-liberal mode of production. This value is achieved through the twin processes of *extraction* and *insertion*. The process of extraction makes the macro-anatomical a priceless site for harvesting raw biomaterials, and as a site for insertion, the macro-anatomical is a locale for witnessing the promised use value itself. While the former is most graphically visible in public and academic debates surrounding the use of embryos for stem cell research, or the harvesting of human organs for transplants, the latter is yet to capture the world's imagination. For example, as yet, there are no scientifically documented instances of embryonic stem cell insertions in macro-anatomical systems. We shall see later how extraction and insertion of embryonic stem cells are being accomplished, and how the resultant question of biovalue is being addressed in India. The importance and centrality of the macro-anatomical in micro-biotech interventions are therefore indispensable, and another good reason to resist Donna Haraway's (1990) call to write the death of the clinic (Bharadwaj *et al.*, 2006).

The generation of biovalue through the process of extraction and insertion is enabled because macro-anatomical sites become bio-available in the first place. Lawrence Cohen's (2004, 2005) formulation of bio-availability captures the process through which the selective disaggregation of one's cells or tissues, and their reincorporation into another body, are made possible. In addition to Lawrence's usage, bio-availability, in the realm of medicine, is taken to mean the degree to which a drug or other substance can be absorbed and utilized by those parts of the body on which it is intended to have an effect. In other words, it is the proportion of a drug which reaches its site of pharmacological activity when introduced into the body. Taken together, these two different meanings of bio-availability bring into focus topographies of bio-available transfers of cellular and tissue-based entities, mediated by technical and biological processes.

These processes either facilitate or stifle the absorption/utilization by those parts of the body on which such transferred biogenetic entities are intended to have an effect. These transfers, achieved through extraction and insertion, and administered as an intended medical resolution of a pre-existing social or medical problem, may be termed *bio-crossing* (Bharadwaj, 2008). This is a crossing across the borders of biology, between biology and machine, and across geo-political, commercial, ethical and moral borders. 'Bio' can be visualized as an ethnoscientific rendition of the 'human biological' and, as such, stands without necessarily being opposed to multiple cross-cultural conceptualizations of human biology. Thus bio can be gainfully conceptualized as an instance of *biologically based biography*, be it individual (for example, an illness narrative or a cultural/'ethno' conception of human body) or institutional (for example, bio-science/medicine/technology). A biography is inextricably inserted in any individual or institutional understanding of the biological.

The anthropological and ethnographic endeavour, in this respect, is to decode the biographical inscriptions that produce, sustain, and alter the bio through multiple crossings across various borders and thresholds. This makes bio a site in which the discipline of anthropology, science, bio-available 'patients', and subaltern subjectivities, are all equally implicated. Bio-crossing, in this respect, is a crossing made in a contested terrain. Anthropological assertions about imploding boundaries between nature, culture, biology, and society need to be problematized alongside the scientific dogma and the everyday negotiating practices of those who find that their bodies double up as sites for both an enactment of new scientific applications and their anthropological narration. Bio, therefore, can never have a stable conceptual mooring, not because it is continually remade, but, rather, because it is 'crossed over' between assemblages of different cultural terrains (cf. Ong and Collier, 2005). The notion of an assemblage when invoked as dispersed in the Marxist usage purports to an idea of use value that may be looked at from the point of view of 'quantity' and 'quality' of any useful 'thing' (in the Marxist formulation these were things like iron, paper, etc. though in the present case substituted by the idea of the [bio]logical). Thus, 'it' (the useful thing) is:

> [an] *assemblage of many properties*, and may therefore be of *use* in various ways [extraction, insertion, etc.]. To discover the various uses of things [such as the very idea of bio and its use value] is the work of history. So also is the establishment of socially-recognised standards of measure [e. g. bio-medicine/technology] for the quantities [such as bodies and bodily tissues] of these useful objects. The diversity of these measures has its origin partly in the *diverse nature of the objects* [e.g. the nature of an object will amount to its discursive contours such as bio, body, 'ethno' physiology] to be measured, partly in *convention*.
>
> (Marx, 1963: 43, emphasis added)

Thus, an assemblage has both historic depth and future potential; it actualizes the past and future into the present, not so much through severing all connections from temporality, but by incorporating such links into the shifting use value of a given 'thing' by (re)measuring it in the present using the 'conventional' or normative standards of the day. This value, produced through a multitude of 'qualities' and 'quantities', is the single most important signifier of an assemblage. Bio-crossing, therefore, amounts to a contextual, contingent and temporal movement of the human body (itself a cultural, historic and political assemblage) across tangible, material, philosophical, historical, political, and many more 'fictional' borders of varying scale.

When actors in stem cell clinics and laboratories are described as bio-available and not biosocial, they are essentially being described as undertaking a bio-crossing emerging from two different readings of the term bio-available. The individuals in Indian stem cell clinical sites emerge not only as bio-available, in that they can be used for extraction and insertion for the generation of biovalue, but also because the ways in which their failing biology absorbs, processes, rejects, or accepts stem cells is highly unpredictable. While they are bio-available to science (predicated on the rhetoric of pre-treatment counselling, informed consent and choice) for generating exchange and use value, that which is bio-available to their biology, in terms of actual tangible benefits, is highly variable.

Bio-crossing 'emerging life forms'

How might we conceptualize stem cell development and application in a mobile world? As an 'emerging life form', stem cells have recently been viewed as 'global biological' entities, distinct from Lock's categorization of 'local biological' or 'local biologies' that emerge from debates about menopause, brain death, and organ transplant (Franklin, 2005). It is at the confluence of such local debates and global discourses about new life forms that bio-crossings are at their most prominent and visible. These are multiple crossings in pursuit of biological knowledge and markets, of health and healing, spread across disparate borders of varying scale. Such crossings force discourses and debates between the local and global conceptualizations of life and its bureaucratic management, of ethics and regulation, as well as debates about the standardization of manufacturing processes, or legislative controls into new alignments, alliances and alterations.

A recent example of an emergent bio-crossing between local debates and global discourses about embryonic stem cell development and application is localizable in a privately owned clinic in New Delhi. A small research laboratory and 20-bed hospital (Nu Tech Mediworld), located on the edge of a semi-urbanized village subsumed into the rapidly expanding metropolis, has become the unlikely epicentre of local and global bio-crossings. The facility is owned and run by Dr Geeta Shroff, who is supported by a small team of professionals including an anaesthetist, a physician, a pathologist and the nursing staff. Dr Shroff graduated from Delhi University's School of Medical Sciences in 1993 and went on to work for a period of time in India's top medical institute, the All India Institute of Medical Sciences (AIIMS). In 1996, she launched her private hospital that gradually evolved into a genetic research laboratory.

The beginnings of Nu Tech Mediworld were, however, very different. Conceived as an infertility centre, with a fully-fledged in vitro fertilization (IVF) laboratory and a small operating theatre, it pioneered a successful intra-uterine insemination (IUI) process with a claimed success rate of 70 per cent. However, more dramatic breakthroughs were to emerge in the new century. In 2001, as the scientific director of Nu Tech Mediworld, Dr Shroff expanded into the area of embryonic stem cell research. However, no-one could have predicted that, in a space of two years, her ESC clinical research would have become the focus of national and international criticism and scrutiny. Part of the reason it did lay with the 'dis-location' of the facility itself. A modest establishment in a semi-urbanized village, surrounded by workshops, small home-run businesses, sauntering cows, rubble and urban pollution, Nu Tech Mediworld appeared a very unlikely site for cutting-edge science to be conducted according to Good Manufacturing Practice (GMP), as required by both EU regulations and the UK's newly created Stem Cell Bank. That this dis-location also became the centre of local and global bio-crossings compounded the disapproval.

Dr Shroff was first contacted in 2004, and through the course of ethnographic observations in her facility, twenty patients were followed who self-identified as responding to embryonic stem cell insertions for conditions as diverse as motor neuron disease, heart conditions, and spinal damage and paralysis. The media outrage in India and around the world focused on the ethics of human trials at a time when, globally, the science of ESC was in its infancy. The dis-location of Dr Shroff's stem cell insertions was compounded because her practice, and, indeed, scientific claims, did not conform to the normative expectations implicit in the work of high investment biotechnology research. For example, Rabinow and his collaborators argue that:

> Within the contemporary human sciences (as opposed to journalistic accounts, in which heroic geniuses make discoveries that everyone else recognizes and hails as true), it is commonplace that science is a collective activity: of individuals spatially assembled in 'a lab'; of large groups of specialists doing parallel work, who constitute competitors and peers; and of extended networks of diverse sorts, including fiscal, legal, political, and now ethical.
>
> (Rabinow and Dan-Cohen, 2005: 63)

This fascinating, albeit Euro-centric, reading of 'baroque science' adequately explains why dis-locations such as Dr Shroff's private research laboratory would be conceived as an anomalous, maverick space. In a global order preoccupied with 'high stakes' science, bankrolled by venture capital and state subsidies, and responding to legal, ethical and moral concerns with a distinctive Euro-American provenance, an establishment such as Nu Tech Mediworld is not surprisingly seen as a throwback to another era, where lone, maverick scientists claimed miraculous breakthroughs in pursuit of knowledge. Dr Shroff's claim, however, is more modest. She used one left-over embryo from her own IVF programme to create the stem cell lines that she now injects into her patients. She contends that she is no genius, and that it was her interest in IVF that caused her to stumble upon a breakthrough that enabled her to culture, cultivate, extract, and insert embryonic stem cells with dramatic therapeutic outcomes. She describes how, as an individual, she occasionally finds her research and clinical application daunting:

> As an individual person who has achieved a lot, you're a little scared because it is so new, well, not any more for me, but you don't know what the response is going to be. You don't know who to approach, what it entails, you don't have government backing, financial support. Luckily, I had the equipment because I am an IVF specialist and the other stuff you just save and buy.
>
> (Interview material)

Dr Shroff's contribution to embryonic stem cell research appears a far cry from the industrial scale of research and development around the globe. In her laboratory, no bigger than a small room, Dr Shroff has made every effort to recreate a GMP standard environment, that is to say a temperature-controlled environment with constant particulate matter monitoring. GMP is an assurance that products are consistently produced and controlled according to the quality standards appropriate to their intended use. Dr Shroff's laboratory has a particulate matter monitored environment, a seven-door entry system, two hepa filters, UV radiation devices, and equipment ranging from freezers to a micromanipulator. A further adjoining split level unit is reserved for isolating and growing cells, with a dedicated area to ready injections for transplantation.

The bio-crossing between each of these reserved physical spaces is carefully monitored, and the environments are graded to maintain optimal sterility. However, as with Dr Shroff's clinical application and science, her cultivation and transfer methods are not validated. Validation criteria are notoriously difficult to define and arrange. For instance, what grade environment is adequate for stem cell extraction or embryonic manipulation? What is an appropriate measure of 'quality', or, for that matter, just how much data is enough? While these issues are well documented and are laid down in the UK Stem Cell Bank's (UKSCB) validation protocol, this cannot be replicated or used as guide to re-enacting similar conditions. In accordance with the UKSCB directive, anyone wishing to upgrade to a GMP 'gold standard' stem cell facility (such as an IVF laboratory) must develop a protocol based on a body of accumulated data and evidence that validates the quality of the laboratory's environment. Such a protocol covers everything from the materials and modes of designing and building physical structures, through the air conditioning, cleaning agents and solvents used, to the use of builders and architects with a demonstrable track record in constructing such structures. For biological material to freely develop and circulate, such validated GMP environments are the only zones of distinction.

According to Timmermans and Berg:

> The gold standard represents the ultimate standard in medicine ... in medicine, the gold standard is regularly used to describe definitive and decisive standards ... it is a measure against which everything else will be measured: it constitutes the rock bottom to which new candidates for standards are compared, and it defines the truth. Gold standards, therefore, do not seem to evolve. Once they are put in place, their authority is so overwhelming that it looks like as if they will resist time.
>
> (Timmermans and Berg, 2003: 26–7)

As their series of case studies show, however, gold standards are locally and contingently produced. Despite their appearance, they depend on professionals' interpretations and judgements. They are not the impersonal and immutable phenomena that they are normally taken for. Consequently, the GMP standard is fluid and contingent in its make-up.

The pursuit of local or global standardization ironically fails at this critical juncture, as the only certainty or 'constant standard' remains the process itself. That is, an individuated validation process based on accumulated data and evidence over a period of time. There is never a standard pattern that can be exactly replicated (Collins, 1985). The local/global, visible/invisible, constant/variable, physical/atmospheric, pure/impure, explicit/tacit circulate in curious new permutations to solidify into a GMP standard, until such time that more accumulating evidence compels further rearrangement. The standard is thus derived from the constant mix and between different domains to achieve an unproblematic bio-crossing. This is not about creating global manufacturing standards, but, rather, it is about creating ideological apparatuses that allow the choreography of designing space and practice to be enacted in the pursuit of a good standard. The norm of sterile purity, therefore, remains in a constant, contingent flux (Douglas, 2005), something very much in evidence at the UKSCB (Stephens *et al.*, 2008).

The idea supersedes the norm. If Dr Shroff can show enough accumulated data as evidence of a constant and standardized environmental control that delivers clean stem lines, she would have, in effect, met the UKSCB requirement. However, her contention is more complex. She questions the scientific rationale behind maintaining such optimal levels of sterility (grade A and B environments in GMP parlance). Clinical application as a matter of necessity requires, according to her, that cellular transfers take place in highly contagious operating theatres (or Grade D environments). How the bio-available body accepts or rejects such cellular transfers remains another fascinating issue, as the bio-crossing is made from pristine (laboratories) to highly polluted (operating theatres) biological environments.

From biosociality to bio-crossings

In the global age of stem cells, the really tantalizing potential for cures emerges alongside the real danger of human exploitation. In several cases, patients from all over India, and from as far away as Britain and the United States, undertook bio-crossings to New Delhi to try stem cells, their only hope and as the medicine of last resort. These individuals and their accompanying family members could be viewed as acquiring 'biological citizenship', in that they may be seen as existing in the 'political economy of hope' (Rose and Novas, 2005: 442). Faced with life-threatening conditions whilst their biosociality (Rabinow, 1996) is not, as yet, fully expressed, they remain unable to articulate the 'doubled discourse' of acceptance and normalization (Rapp, 2000). We can only speculate that perhaps one day they will become bio-socially organized (Bharadwaj, 2006a). However, a deeper examination of the biological biographies that shape the everyday struggles of patients reveals a different reality. In all recorded encounters, patients, who were seeking embryonic stem cells as therapies for conditions ranging from infertility to a complex and diverse pool of degenerative diseases, were focused on the question of resolution, relief, rescue, redemption, and restitution, rather than biosocial mobilization. They have a complex investment in the therapies offered by Dr Shroff that, in many cases, result in dramatic outcomes. They are, nevertheless, not assembling around their failing biology with the view to actively carving out a sociality. Unlike her critics, who in the absence of any credible animal studies, data, and peer-reviewed research, continue to attack Dr Shroff as a dangerous maverick, her patients view her as 'worth a try' so long as she can offer relief, however temporary.

Dr Shroff's patients are drawn from a cross-section of local and global individuals and their families. Her very first bio-crossing, however, became possible because of her parents. She treated them with embryonic stem cells for two very different heart and skin conditions. Her next patients were her mother and father in-law, and she has since gone on to treat 200 cases from within India and abroad. Selection criteria remain focused on terminally ill cases, and on individuals with severe disabilities who have no hope for a cure or improvement to quality of life. Patient backgrounds range from humble rural peasantry, educated professionals, civil servants, bureaucrats, and a central government minister, to a range of non-resident Indians and foreign nationals from as far away as Australia and the United States. Interestingly, according to Geeta Shroff, 25 per cent of her 200 patients are either medical doctors or their relatives.

According to the ICMR Guidelines; '[T]rials in human patients will commence only on those patients where no other form of treatment is available and where, in the absence of the transplant, the patient is likely to suffer relentless deterioration in his health with fatal termination' (ICMR, 2000). Dr Shroff states categorically that her embryonic stem cell therapy, in its present stage of development, is being applied to improve the quality of life of the terminally ill, their families and relatives. She contends, 'in the future stem cells will cure, now we're using them to improve quality of life'. Such dramatic admissions have been reported in national and international media, including the UK *Guardian* newspaper and the Sky News network. In these reports, Dr Shroff has consistently maintained that she is possibly the only scientist in the world today who has made a successful transition from extraction to insertion using stem cells, with demonstrable biovalue in the (shape of use value) as her only validation and proof (see Chapter 5). Despite the domestic and international media casting aspersions on her methods and technique, and indeed casting her in the predefined and available category of the maverick, anomalous scientist, Dr Shroff remains focused on her patients, who have in turn developed an almost devotional attachment to her, their saviour. Even the most sceptical of media reports could not overlook the fact that fathers were breaking down in tears at the sight of their paralysed children take their first tentative steps after eleven or more years without walking, or the story of Jassy, an affluent London-based NRI (non-resident Indian), suffering from rapidly advancing motor neuron disease, who is swallowing food and breathing without assistance a year after her bio-crossing to India.

The case of Jassy provides a good example of how bio-crossings can dislocate local/global spaces, ethical/legal distinctions, and most importantly the research/application protocols. An excerpt from Bharadwaj's field diary records some revealing encounters with Jassy, her family and Dr Shroff:

I first heard of Jassy in late autumn of 2004 in Cardiff. Unexpectedly the phone rang and the male voice proceeded to introduce himself as Jassy's son. 'My mom has Motor Neuron Disease, we are thinking of going to Delhi to try Dr Shroff's stem cells'. Apparently Dr Shroff had given my contact details to his family and the son was now interested to know what I thought of the procedure and whether or not I would advise them to proceed with the treatment abroad. 'Can't say you should or should not', I heard myself saying. 'The procedure is experimental in nature.' I further added, 'I personally don't know what you should do.' The conversation continued and he repeatedly asked technical questions about the procedure and what it entailed. I could not have been more ill prepared for this call but it did set into motion a series of meetings with Jassy and her family. My second contact with Jassy was in Delhi in the summer of 2005. She had flown in with her son for embryonic stem cell treatment and the day I first met her she was recovering in a little private room. She looked cheerful and gave a broad smile, saying, 'I am much better.' I couldn't get much more from her but three weeks on, days away from their London departure, I spoke to her again and whilst her speech was slurred she looked visibly happy and pleased, she exclaimed, 'Breathing is good!' By autumn 2005, Dr Shroff was in London, en route to the USA to present her ESC research results at the World Congress of Cardio Thoracic Surgery. We arranged to meet near her central London hotel. She pulled up in a BMW with a woman at the wheel, 'jump in we can't park here, quick!' The car sped past the crowed end of Oxford Street past Marble Arch as we said our hellos. 'We are going to see Jassy', she said, 'this is her sister.' A 50-minute run through the crowded city brought us to the leafy edges of London and finally to an imposing house where Jassy lived. Her excited family and friends were gathered in the drawing room to meet the doctor from India. My unannounced and somewhat awkward presence was explained by Geeta, 'He is researching stem cells in India and I thought he might like to know how Jassy is getting on, you met him in Delhi, remember?' Jassy flashed the now familiar smile and began to speak. She explained how her speech was improving and she could now swallow and breathe a lot more easily. 'You are due back in India soon and we'll give you a top-up dose.' Dr Shroff added, 'It's a shame I can't give you the cells here and now.' 'Can you do that?', I asked. 'Yes, of course, I have made the cells very portable and mobile. If I wanted, I could have easily carried them packed in an ice flask onboard in my hand luggage. I could then inject her in the comfort of her home without her having to fly to India ... But that, you see, would be breaking the local laws, I can't do that legally here, so Jassy is going to have to see me in Delhi again for her top-up.'

(AB field diary)

Dr Shroff further explained how stem cells could be made portable like antibiotics, with an adequate shelf-life if stored under proper conditions, and thus able to move around the globe and made accessible. The bio-crossing Dr Shroff was suggesting has clearly been in the realm of fiction, but now seemingly realized (in her and her patient's world), defying all 'rational' thought process. This proposed bio-crossing was not only beyond policed borders, but it is also in a legal, ethical and scientific dis-location. Her critics, both local and global, have continued to attack her methods and claims as deeply unethical and dangerous. Dr Shroff, however, has hit back defending her technique and questioning what might bio-ethics look like in a globalized research system:

What is ethical practice? Ethical practice is one in which you do every-thing to save a patient. A practice with best of intentions and is geared to save life, that is if you can save a life to the best of your knowledge.

(Interview material)

Responding to the federal ban on embryonic stem cell research, Dr Shroff lambasted the US presidential decree of August 2001 (see Chapters 4 and 6).

She criticized nation–states that sign up to international regulatory regimes in order to secure great international recognition and impact for their scientific innovations. Criticizing the moves to bring the existing ICMR Guidelines in line with the Euro-American concern about ethical and regulated development of embryonic stem cells, she contended:

> [W]e here in India ape the West, that's just one problem … we follow the Indian council of medical research guidelines. They [the ICMR] have looked at the HFEA and we here don't have any problems with the embryo [being] a human being … we are very practical about it … how can few cells going down the drain and put to better use [be unethical]? Is unborn life more important or living, suffering people? As far as technology is concerned, we have indigenised it, we follow the textbooks but we have to readjust these technologies to suit our patients. Abroad you won't even think about TB [tuberculosis], here I am thinking 'Does she have TB?' Official circles [e.g. the ICMR] don't think, they pick up [ethical guidelines] from here and there, no original thinking, original thinkers are crucified and then there is the pressure from outside, stem cells is a major issue. The R&D is not as developed in India as in the West. Stem cells are a regenerative medicine, medicine of the future. Pharma companies, why would they want breakthroughs? India is one of the oldest civilisations and as long as we continue to be Indian, there will be no problem but the moment we start aping the west, the minute we do not use our brains, that's when the problem starts … why should one size fit all? Are they always right? Did they find weapons of mass destruction in Iraq? And yet a country was attacked and millions were put through suffering, because they were right? The ethics of humanity ought to be there. If they cannot manage the ethics of humanity, who are they to say what is biomedicine, what is right and wrong? … I say, learn the best from the West and reject the rest! They should do the same, learn the best from the East and reject the rest!
>
> (Interview material)

This sentiment was echoed by a lot of fertility experts and stem cell scientists. A leading fertility expert in Mumbai (Dr X) commented:

DR X: They don't mind killing thousands of Iraqis; that doesn't bother them. They don't mind Iraqis being shown live, nude on television; that is not considered cruelty. Only what suits them is considered cruelty. I think we should not have double standards in our lives.

AB: I was looking at the legislation in Germany, for instance, you cannot work on embryos or stem cells but …

DR X: I have so many requests from Germany.

AB: It's quite interesting in that sense.

DR X: It's a double standard, no?

AB: You don't tend to get that here in India. What is it that makes India so different?

DR X: I think it's our culture, our religion. Religion has always taught us to be extremely tolerable to everybody else and other religions. So it's a sense of bonding, it's a sense that something exciting, something new is done. We don't have people waiting in the curtains trying to shoot down something, like, the stem thing started because of some political reasons, OK? What Bush has done to America with their embryonic stem cells has really pushed it a decade back, more than a decade back.

    The potential is a lot. We have ethical issues but we don't have baggage to carry. We have our ethics committees in place, we have a scientific advisory committee, but we don't have a lot of political baggage, at least not yet. Maybe because people – What we should do is we should see what is important for our country. We should not ape the West in trying to do fancy things like genetic profiling. Our strengths lie in the industrial background, in our plant biotech, in our ethical ethnic biodiversity. So we should try and do things in biotechnology which will have direct offshoots which we will be able to reap the benefits (*sic*). Also whatever laws, whatever ethics that we want to follow, we should do in concordance to what is relevant to India. You don't have to follow what the West says. Like, if the West says to transfer only one embryo, I'm not going to transfer only one embryo, I will make sure that I transfer four embryos. I will take the consequences of multiple gestation. I will do that not because I want meta-pregnancies but because I know that on an average, Indian couples cannot afford, they will do on an average 1.2 cycles. If I'm going to give the best to my patient, I cannot put in one embryo and freeze the remaining. So what applies to, say, Germany or France where they get three cycles does not apply to my patient population. We shouldn't buckle under pressure and not try to ape anybody in setting our guidelines.

(Interview material)

Through the course of this research, scientists, researchers, patients and donors alike pointed out a number of local and global issues that, to them, set aside the unique character of Indian culture and science. While the dislocation of stem cell insertions that Dr Shroff was spearheading in her small facility in New Delhi was a far cry from larger global developments in research, she nevertheless espoused an ethic that could only work if political and cultural borders between nation–states were transgressed to create contextual and contingent ethical models. The global media interest in her stem cell therapies, and the barrage of criticism from scientists and regulating authorities, replicated, in microcosm, the state of the world where unjust wars could be waged on sovereign nation–states. The ethic of humanity, of offering health and healing, was uppermost in her proposed ethical world-view. It was the end that justified the means, even if that meant digressing from the established scientific modality and bioethical thinking that, in the context of the Iraq war, had become entrenched as 'western dogma and double standards' in the popular imagination. Promoting bioethics in a dis-located (and now disenchanted) world somehow equals militarily promoting democracy in disparate corners of the world where 'truant states' resist such external 'ideological impositions'.

The debates surrounding moves by the ICMR to legislate stem cell research in India, and moves by the Indian state to adapt governance models to facilitate global participation in the burgeoning global moral economy (see Chapter 6), have only exacerbated the disquiet among sections of Indian scientific community that saw in these moves a cynical ploy to curtail and discredit smaller research laboratories as 'not up to international standards'. Dr Shroff's project, therefore, is not one of creating an 'archaic modernity', or some kind of pseudo-nationalism that is inward-looking and protectionist, but rather is one of creating a bio-technical intervention which is crossed between multiple domains of ideas, practices, ethics and culturally-based common sense. Dr Shroff is not rejecting participation in the global political economy of biotechnology, and her moves to acquire intellectual property rights on her breakthrough is a clear indication of this, but her refusal to conform to scientific norms is focused on a rejection of the moral economy of stem cell generation and circulation around the globe. She is a renegade to her detractors, a shadowy maverick dangerously flirting with science. While an identity ascription such as this can, and is, being vocally resisted by her, it cannot be recapitulated under the banner of biosociality. From patients suf-fering the biological consequences of a spectrum of disorders, to margin-alized clinicians like Geeta Shroff, biosocial mobilization is almost never identity-forming. Their survival strategies are focused on managing the con-sequences of identities that are almost always ascribed as opposed to achieved or assumed.

Dr Shroff, however, is no maverick to her patients. In actual fact, she is a woman doing a science dominated by men, an Indian in the world led by the United States, and a human with real material ambitions. The reason her breakthrough has not been adequately peer reviewed and validated by the international community is primarily because she finds the process of trans-cending from having successfully generated use value to creating exchange value very difficult. As a private individual, warding off domestic and inter-national pressure, with her pool of 200 bio-available patients who in turn have all but surrendered their bodies to her scientific breakthrough, her only remaining concern is how to protect her intellectual property in a globalized research system. Rightly or wrongly, she fears the scientific world will be quick to discredit her and walk away with her scientific gains. In other words, there are good commercial reasons why she continues to be sceptical about sharing her breakthrough. While she is pressing ahead with acquiring local and global patents to protect her breakthrough before making public her extraction and culturing method in peer-reviewed scientific journals, these moves are only a partial safeguard against fears of intellectual property theft in a globalized research system.

The UK newspaper *Independent on Sunday* carried a story on 29 January 2006 explaining how technology firms increasingly fear that, in applying for patents, they unwittingly give away their secrets and profits to rivals. The time lag between the publishing of the report on the patent application and the actual granting of the patent opens a window of opportunity that enables competitors to take advantage. Even in the best case scenario, an element of risk cannot be ruled out. These fears notwithstanding, as the local and global chorus of disapproval gathers pace, and international calls for better state control and regulation in India gather momentum (*BMJ*, 2001), it is very probable Dr Shroff and her patients might be quarantined by the Indian state.

The Indian government appears busy rushing new legislation through parliament with the view, at least in principle, to better control the research and application of embryonic stem cells in the country. Once these laws come into effect, which will not happen in the foreseeable future (see Chapter 6), it is likely Dr Shroff will either be forced to shut down her clinic, or be barred from recruiting any new patients. Dr Shroff, however, is fighting back and is finally taking steps to publish her research findings and to protect her intellectual property by patenting her stem cell extraction and insertion technique she pioneered in her small research lab in Delhi. Perhaps there will be a biosocial rising, as her patients and similar others demand access to a potentially 'dangerous' and experimental stem cell therapy, which, for the moment, seems to work. However, informal conversations with many patients and their families indicate that they do not so much predicate their discourse on their biological biographies, but rather around a woman who is able to offer the promise of health, hope and healing. Therefore, should there be a biosocial mobilization, actors will rise and congregate around a 'maverick' woman scientist, rather than a mere scientist in a white coat, a *devi* or goddess to many, who bestows the gift of life.

There are early signs that Dr Shroff is already being seen as a *devi*, which opens a Pandora's box of cultural analysis and gendered explanations. However, even in the event of such an uprising, it is highly problematic to conceptualize charismatic authority and beseeching supplication from patients as an instance of biosociality. As argued elsewhere (Bharadwaj, 2006b), it is not uncommon for science and religion to forge a symbiotic relationship in Indian IVF laboratories, and for physicians and clinicians to be viewed and revered as life-giving, sustaining and destroying deities. In the IVF clinical and laboratory spaces, it is fairly routine for Indian scientists and clinicians to display overt religiosity, and to enlist divine assistance, in the process of high-tech conception. Unlike the Euro-American 'post-moderns', the Indian scientists, and patients, turn to techno-science to overcome their fate, while continuing to rail against the unyielding 'gods' (Rabinow, 1996: 103). The concept of biosociality thus fails to capture the place of the *supernatural*, both as cultural practice, and as the experience of the 'uncanny', 'pre-objective' unknown (Throop, 2005). Furthermore, the very act of petitioning the state, and the resultant politicization of the individuals, bypasses the biosociality question as well. Demands for relief and access to resources, by drawing attention to the pain as 'the price of belonging to a society' – especially where such demands challenge the certainty of bureaucratically defined medical science – have, in the past, only transformed victims/patients/sufferers in India into mere 'malingerers' (Das, 1995: 137–74), as opposed to biosocial actors.

Conclusion

For very different and difficult reasons, stem cell clinical trial patients in India have endured bio-crossings, extractions and insertions. In so doing they remain biosocially inactive, inert, indifferent. This is hardly surprising, because so long as we live in a world where the poor, the unfit, the gendered, the stigmatized – to name a few problematic categories – are not so much killed as 'allowed to die' (Das and Poole, 2004), concepts like biosociality will have to work very hard to explain both the idea of bio and the nature of sociality. This is not to claim that biosociality does not exist in India, but rather that the social trajectory of the bio, and its biographical inscription, seldom produce individual or group identities. Even in the case of scientists and clinicians such as Dr Shroff, who are vocally critical of the proposed regulatory moves in India, and the global media and scientific denouncements of maverick science, the resultant biosociality is never quite identity-forming. Like the stigmatized fertility seekers, actors like Geeta Shroff often have their social identity ascribed to them – in this case, one of a reviled maverick. Perhaps it is a feature of her dis-location where, not unlike those who undertake bio-crossings for extractions and insertions, both science and the scientist remain outside ideological and normative validation.

The chapter has argued that the relatively poor and sick, as well as the prosperous but sick, bodies in Indian stem cell labs are seldom biosocial, but, rather, are always bio-available for bio-crossings. To overlook this is to overlook the social and health inequalities across cultures, and the differing opportunities for action and expression that such inequalities engender. The poor in a neo-liberal world inhabit biologies that, following Nancy Scheper-Hughes (2000), are routinely construed as waste. It may be further argued that such waste gains value only when it can be recycled in a tangible, meaningful and, above all, profitable sense. Thompson argues that in the 'biomedical mode of reproduction', unlike the capitalist mode of production, waste is seldom a political or logistical problem but rather an ethical one of how to designate life material, such as embryos, as waste (Thompson, 2005: 264). In the neo-liberal mode of production, this problem is frequently addressed by resorting to recycling socially/ethically defined waste or surplus that cannot otherwise be gainfully accumulated. Through modern transplant technology, the biosociality of a few, argues Scheper-Hughes (2000), is made possible through the literal incorporation of the body parts of those who often have no social destiny other than premature death. The same argument can be made for gendered and other kinds of marked bodies, which die real and social deaths as these bodies are deemed unimportant, unsalvageable or simply substitutable. A bio-crossing, therefore, is a crossing made in a social space fraught with opportunity and danger which, on occasions, can be a calculated risk or a forced dislocation as the last resort.

In the light of the foregoing, how might the issue of agency and its relationship to structure be better understood? While individuals and collectivities (as numerous anthropological and sociological studies have shown) seldom passively submit to the brutalizing effects of power and knowledge, the ethnographic disquiet, in the context of India, must always centre on the question: what, if at all, is politically achieved from (often passively) resisting? For example, the model of passive, suffering Indian womanhood, or the Hindu ideology of *pativrata* (literally, she who takes a vow [*vrat*] of devotion to her husband [*pati*]) (Harlan and Courtright, 1995: 8), was clearly instilled in the ancient religious laws of *Manu*, which still hold considerable sway over Indian women (Dhruvaranjan, 1989). While such an ideological conformity might be the only available resource for some women trying to survive, negotiate, bargain, compromise and secure concessions, dispensations, extensions or exemptions, it nevertheless leaves the patriarchal structure unaltered. However, resistance – of the brutalized, the weak and the infirm – can alter the structure at some rudimentary level, by remaking the conceptual boundaries of institutions such as the state through the very acts of securing survival and seeking justice from the margins (Das and Poole, 2004). That said, such altered structures seldom lose their power to code certain kinds of knowledge about individuals, populations and incompliant bodies, and seldom tolerate the dilution of their ability to police, patrol and punish transgressions of dominant cultural norms and ideas. One example can be found in the issue of identities as rooted in agency. It is an anthropological truism that identities are multiple, complex, contextual, contingent and sometimes consensual. As an identity-forming resource, biosociality enables people to congregate around some aspect of their (failing) biology. In the everyday, local, moral worlds of patients and their carers, however, it is not so much identity issues that frame actions, but rather life shaped by demands for resources and treatment, a cry for help, or a relentless but exhausting search for resolution. In the Indian context, this entails forging tactical alliances within and outside families, as well as alliances borne out of the everyday emotional work of care-giving (Bharadwaj, 2003). According to Das (2001), alignments between family and state embody a 'politics of domesticity', which involves connected body selves, and not liberal biosocial individuals in India (Das, in Roberts, 2007: 82). More recently, Das and

Addlakha (2007: 147) have argued for more ethnographies to evaluate whether biosociality is to be understood primarily in terms of affiliative, associative communities, or whether post-modern forms of sociality demand a reimagining of affiliative community as well.

Equally, Das also 'disallows any scope for entertaining the idea that the family and the community are more protective of the self than other collectivities ... [since] finely woven structures of power within the family often determine the level at which ill health of different members will be tolerated' (Das, 1990: 43–5). In other words, there are those who suffer in silence but show resolve, or openly resist with great tenacity, but never develop a sense of identity other than one that is socially received or ascribed (for example, the inauspicious barren woman, the emasculated infertile man, the senile, the invalid, the leper). These identity ascriptions may, in turn, be negotiated, or resisted individually, or resisted through a network of familial kin, but seldom challenged to a point of total rejection.

The biosociality question, in other words, falls between two contradictory ideological pulls, one promoting the notion of choice and the other denouncing the very idea that certain kinds of choices should be made. In the neo-liberal order, it is never enough to just choose, but choice must also be seen to be rational. By choosing to embrace an experimental therapy, individuals undergoing stem cell trials in India fundamentally fail the 'rationality test'. Their biosociality remains unfulfilled, despite their ironic bio-availability being the only expression of their curtailed agency, which in turn is actualized by undertaking unauthorized bio-crossings across political, legal and ethical borders.

Battaglia (1995), citing Strathern (1988), contends that the agent or acting subject may be less a locus for relationships than a 'pivot of relationships ... one who from his or her own vantage point acts with another in mind'. Agency thus approached can, perhaps, distil the way in which agents are formed through relationships, exchanges and interactions. The unquestioned submission and (bio)availability of bodies in stem cell trials are examples of active agents working within the limitations set by relationships, resources, ideologies and normative expectations. These struggles point to the definite limits to an idea of agency, as, even in the mode of resistance, the actions, strategies, and subversions can amount to no more than the weaponry wielded by the weak (Scott, 1985). However, power relations and knowledge practices, like the cultures that contain them, do not hold still for their portraits (Clifford and Marcus, 1986). Resistance and critique, however passive, produce eventual change, facilitate recoding of knowledge practices, and realign power relations. That this is seldom achieved without a cost and inevitable social suffering remains, more than ever before, the defining feature of life under the neo-liberal dispensation.

4 Sacrificial gifts

Infertile citizens and the moral economy of embryos

Introduction

In the archaic modernity of *neo*-India, women's reproductive potential has come to be viewed as both a scourge, most graphically illustrated by the aggressive and gendered nature of the population control policies pursued by the state (Van Hollen, 2003), as well as a boon, in that this potential is the reproducer of the Indian state and economy. In great measure, *neo*-India owes its rise to the army of its young workforce, and a reserve pool of a staggering 500 million people under the age of 19. In this respect, by fulfilling their traditional fecund remit within the parameters set by the state, Indian women make their reproductive labour valuable in the neo-liberal mode of production (Bharadwaj, 2006c).

The contentious issue of the ethical sourcing of human embryos for ESC generation has required the Indian state to examine how an ethical, but steady, supply of human embryonic tissue might be achieved to facilitate participation in the global moral economy. In line with the emerging global governance modalities spearheaded by the Euro-American countries, moves to legislate and promote embryonic research have led the Indian state to embrace guidelines that in large part draw inspiration from regulatory frameworks in the United Kingdom and the United States of America. This 'ethical embrace' has renewed the focus on women and, ironically, on their lack of reproductive potential, hitherto a subject confined to the realm of the family, community and religion. Women and their reproductive viability remained at the heart of family planning policies, promoting the two-child norm as the developmental ideal for a modern and prosperous nation, for much of India's post-colonial history. However, in the new century, infertile women and their technologically induced oocytes and embryos are rapidly becoming state subjects in need of regulated development, production and circulation.

This chapter examines how India's participation in the global moral economy lies at the intersection of local and global moral, economic and political concerns. In so doing, the chapter will discuss how the sacrifice of the embryonic form for stem cell research is becoming assimilated in discursive realms as diverse as global, bureaucratic, and moral reasoning, and in the local moral worlds of donors, who not only become enrolled as ancillaries to the embryonic stem cell (ESC) production process, but are also enjoined to become neo-liberal citizens of neo-India.

Outsourcing the moral and other standards

On 9 August 2001, President George W. Bush allowed federal funding of embryonic stem cells research to go forward only on cell lines already in existence. Around the world, only 64 cell lines could meet this criterion, and of these ten existed in India, where three lines were held by the state-owned National Centre for Biological Sciences, Bangalore, and a further seven were under development at the private research laboratory Reliance Life Sciences in Mumbai. In early August 2001, the United States National Institute of Health announced that Reliance Life Sciences, along with the National Centre for Biological Sciences would be among ten institutions world-wide that would benefit from federal funding for stem cell research. Both centres have met the US administration's criteria for derivation of human embryonic stem cell lines.

This case marks a watershed in the global politics of generating embryonic stem lines. A globally dispersed network of research centres, extracting cells from human embryos, and located in diverse political formations, can in theory be federally funded and fed by the United States. This unprecedented move linked the contemporary *biopolitics* in the United States, and its moral and cultural reasoning, to strategic economic and political calculations. Equally, the morally contentious issue of human embryonic form, and its sentient potentiality, could for the first time be outsourced to global locales, where embryos could be purged of their cellular cargo to yield potentially therapeutic and research-worthy stem lines. In one deft move, the US President marked embryos, destroyed prior to his 9 August decree, as profane compared to their sacred counterparts, freezing in a moral liminality of IVF laboratories. These surplus and orphaned embryos could however be thawed for adoption (Genchoff, 2004: 766) to (re)establish the 'as-if' (begotten, genealogical) principle of American adoptive kinship (Modell, 1994).

This kinship with a biogenetic form is not merely nationalistic (because American embryos must not be 'killed') but it is also the religious, moral and sacred duty of the citizens enjoined by the state. In other words, the state of exception invoked by the sovereign, following Giorgio Agamben (1998), is one that sets apart the citizen (American embryos) and the non-citizen 'other', the *homo sacer* who 'may be killed but not sacrificed' (ibid.). The sacred body of the citizen may be sacrificed on the behest of the sovereign and the nation, but the 'bare life' of the *homo sacer* is deemed 'before law', that is, outside the purview of divine as well as human law (ibid.; Das and Poole, 2004). The positioning of certain embryos since August 2001 as before the law and hence 'killable', as opposed to embryos that become viable citizens under the moral and divine law invoked by the state, allows the moral economy of embryonic stem cells to become truly global. A new benchmark has been set. American embryos cannot be killed, but in the extreme and rare case may be sacrificed, as in the case of medical termination of pregnancy, but not in the case of generation of embryonic stem cells. However, what this does allow is a conceptual possibility of positing a moral distinction between the 'sacrificial' and the 'killable'.

Parallel to these developments in the United States, different moves were afoot in the United Kingdom. As discussed in Chapter 6, the United Kingdom established the world's first publicly funded Stem Cell Bank in 2003. Only stem lines adhering to the Bank's stringent ethical guidelines and provenance protocols validating the human embryonic materials became admissible for donation. This development has posed hitherto unexplored questions concerning how the standardization of ethical and regulatory practices might look in a culturally diverse world. Taken together, these global developments provide an insight into the Indian response, and its place in the burgeoning moral economy of embryonic stem cells. For stem lines to be admissible in the Bank, so as to become available for future research and commercial use, the Bank has initiated insistence on certain standardized benchmarks, quality assurance processes and protocols to be adhered which, as described earlier, coalesced into the Good Manufacturing Process (GMP) standard (Stephens *et al.*, 2007).

To remain a global player in this moral economy, India needs to respond to the moral concerns of its research partners and potential consumers of its research end-products. To ensure this, India has begun putting systems in place to meet the GMPs (*Financial Express,* 2005), as well as ethical proto-cols that require informed-consent embryo donations as a basic admissible standard. In a November 2005 press conference, India's Science and Technology Minister described how the Indian government not only sought to strengthen stem cell research in the country by extending accreditation to companies to do research, but also required these companies to follow international guidelines on the GMPs so that their products could be mar-keted globally (ibid.). In addition, access to a viable body of citizens, and their cell-yielding embryos that can be 'sacrificed' to fuel India's participa-tion in the market-driven global moral economy, has gained renewed urgency with such regulatory and production protocols. The generation of informed, autonomous, rational citizens, therefore, becomes a priority in this purportedly zero sum game. Against this backdrop of global developments, the Indian Council of Medical Research (ICMR) recently put together draft Guidelines for stem cell research and regulation (ICMR, 2005), discussed in more detail later in this book. The Guidelines make it mandatory that embryos cannot be produced for the sole purpose of obtaining stem cells, a requirement explicitly spelt out by the UK Stem Cell Bank admission criterion (Glasner, 2005). Should these Guidelines become law, only spare

IVF embryos will be considered legal source of stem cells. The proposal is even stricter than current policy in the United Kingdom, where embryos can, under exceptional circumstances, be legally created and used for the express purpose of research and stem cell extraction.

The move appears even more surprising considering Indian cultural conceptions about life. In India, unlike in the Euro-American context, there is no consensus on the moral status of the human embryo. Different philosophical, religious and ideological persuasions define and debate life in an eclectic and open-ended way (Bharadwaj, 2005b). This ethical discourse is applied in a less critical manner by the Indian state. Indian embryos, like French DNA, seem partially locatable at the 'heterogeneous zone where genomics, bioethics, patients' groups, venture capital, nations, and the state meet' (Rabinow, 1999: 1–5). However, for biotech India, the US presidential decree of 2001 was probably the first time the local/global moral and business sensibilities met and mutated into a third way of conceptualizing previously mundane biogenetic entities. The Indian state and its future biotech investments are locatable in this *moral geography* where the local and the global ethical sensibilities have become curiously 'joined up'. The need to locate embryonic stem cell science in India within the purview of legal and ethical scrutiny must be situated in the concern to manage the increasingly collapsing domains of local and global publics. As already shown, the *biosociality* of Indian publics is yet to attain its full potential (Bharadwaj, 2008). The pre-emptive move is influenced in large part by the Euro-American ethical and governance protocols, which the Indian state appears to have embraced despite the apolitical nature of its embryo research. It is easy to conceive this show of solidarity as yet another instance of the global shaping the local, but the concerns that drive these moves in locales like India appear to be more pragmatic. Ensuring the provenance and strictest ethical scrutiny of embryonic entities does not appear in India to be a local cultural response to the question of life. Rather it is a strategic investment in future global markets, potential scientific collaborators, and probable international lay consumers of embryonic entities and their ethical/moral thresholds.

Sacrificial gifts: the ethics of giving

The language of gifting and altruistic renunciation of human tissues is well established in the biomedical and bioethical discourse (Scheper-Hughes, 2000, 2001; Cohen, 1999, 2001; Waldby and Mitchell, 2006). Social scientists, most notably anthropologists, have variously conceptualized the gifting of human biogenetic and tissue forms. The most notable insights have emerged from studies examining the global transactions in human organs such as kidneys, where as Scheper-Hughes argues, '[the] language of "gift", "donations", "heroic rescue", and "saving lives" masks the extent to which ethically questionable and even illegal means are used to obtain the desired object [kidney]' (2000: 198). She further argues:

[W]hat is different today is that the sacrifice is disguised as a 'gift', a donation, and is unrecognized for what it really is. The sacrifice is rendered invisible by its anonymity, and hidden within the rhetoric of 'life saving' and 'gift giving', two of several transplant 'key words' we are trying to open to a long overdue public discussion.

(Scheper-Hughes, 2001: 54)

The case of 'embryonic gifts', however, is more complex, as it cannot simply be recapitulated as mere gift of life, but rather also 'gift of potential knowledge to a medical researcher' (Waldby and Mitchell, 2006: 70). This produces value, not through mere circulation, but, more significantly, through transferral into a derived cell product, with accompanying claims to intellectual property and ownership detached from the point of embryonic conception. However, as we shall see in the next section, in India the relationship between embryo ownership established through kinship, and that asserted through intellectual property, is being further complicated.

The gifting of spare IVF embryos in India, in line with the prevailing practice in other global locales, has hastened the creation of citizens who, from the point of view of the state and its legislative modalities, can be imagined as imbued with individual rights. They can also be imagined as possessing bodily autonomy which in turn allows them to give well-informed consent from a position of knowledge, and encourages them to make choices reflecting free will. Similar assumptions are implicit in the arguments favouring the open commercialization of organ trade around the globe. Nancy Scheper-Hughes describes how the Bellagio task force – a small international group of transplant surgeons, organ procurement specialists, social scientists, and human rights activists – concluded that:

the sale of body parts is already so widespread that it is not self-evident why solid organs should be excluded [from commercialization]. In many countries, blood, sperm and ovum are sold ... On what grounds may blood or bone be traded on the open market, but not cadaveric kidneys?

(Scheper-Hughes, 2000: 197)

The response, from social scientists and human rights activists serving on the task force, remained profoundly critical of this bioethical discourse predicated on 'Euro-American notions of contract and individual choice'. These critiques laid bare the way in which the 'social and economic contexts ... make the choice to sell a kidney in an urban slum of Calcutta or in a Brazilian *favela* anything but a free and autonomous one' (ibid.).

How might this critical frame help contextualize the deeply cultural and socio-economic issues at stake in securing 'consent, contract and choice' in embryo donations in India? This is especially sensitive when considering those for whom infertility or reproductive disruption on a similar scale is a profoundly disabling condition, especially in the context of classic patriarchy as found in the patriarchal belt extending from North Africa, through the Middle East to southern and eastern Asia (Inhorn, 1996; Inhorn and Bharadwaj, 2007). The virulent stigma attached to reproductive disruption in India is in fact better understood when seen as shaped by *pro-natalist imperatives* (Bharadwaj, 2005a). These can be understood as patriarchal strictures that mandate reproduction, privileging and conflating motherhood with womanhood, and fatherhood with manhood. Unlike fertility where, as Ginsburg and Rapp (1995) argue, the notion of *stratified reproduction* helps clarify how 'some reproductive futures are valued while others are despised', infertility in India poses a different set of questions, where the patriarchal normative and ideological order has, for centuries, had a vested interest in the reproductive futures of its populations. Thus, in neo-India today, as in the past, very often only one kind of reproductive future is despised, namely a future obstructed by infertility.

Since the biotechnology of embryonic stem cells has become thinkable, the 'stifled potentiality' of reproduction has attracted both public and private sector attention in India. This is most graphically illustrated by the proposed ICMR-sponsored ethical guidelines on infertility management and embryonic procurement. In fact, moves to monitor and regulate the steady growth in the number of IVF clinics, as well as egg and sperm donation in private sector practices, has created a 'furore' in the medical community in India (Express Healthcare Management, 2000). The main bone of contention within the Indian Council for Medical Research guidelines is the proposed prohibition of intra-familial gamete donation. The medical community has reportedly taken strong exception to the ban on sperm donation by a relative or known friend of the wife or the husband, fearing that this would trigger paid donation and trade in semen. Both fertility experts and stem cell research scientists interviewed for this research vocally opposed this legislative move. An eminent Mumbai-based fertility expert, Dr Sarika, summed up the mood:

> It's not at all realistic because these guidelines will only help people to do things without the arm of the law. If people are stopped, they'll go and look round and do it. So that part of the ethical guidelines is not making sense and I have been part of the drafting committee, so we have been telling them that this will not make sense ... It looks like a cut and paste job, yes, it does. I think that part hasn't been really thought about. We haven't ruminated enough on that, although all the gynaecologists on our part have made it very clear that we definitely need to have people coming forward who want to use their sister's, cousin's, friends' [gametes]. It is a guideline but most clinics will not follow it because what if anyone who says 'I am not related'? She is a donor. There are ways of getting one. ... How? Why should you? ... I think the doctor's role should be very well defined and limited to her area of expertise, as sometimes I tell my staff people that when there are social issues we

should stay clear of them. We can guide them to certain things. When you realize that there is some reason for infertility and it's becoming now a social issue where there are so many other factors, one should really stay clear, because then you know you are impinging on somebody's personality. So I don't think this is really in the guidelines ... It is going to go to parliament at some point or the other. At some point in time it will become law, it will. I don't know in how many years but the idea is that it will become a law ... We have had several meetings with ICMR people. We have told them, we have expressed ourselves and so have other gynaecologists, but after the last meeting again, which was almost a year ago, it's not moved forward. You know what happens sometimes, the government changes and the bureaucrats change and then you have to restart the whole thing again. We are at that point in time that although we have very strongly said that this is not going to work, because of certain changes, we've not been able to have a next meeting.

(Interview material)

Should the Guidelines become law, the consequences of such moves may impact on family-forming strategies employed within the confines of clinical spaces. Here, tactical alliances are forged between select family members, infertile couples and their clinicians (contrary to Dr Sarika's foregoing assertion) to keep donor gamete IVF secret, and, on occasions, as close to the 'agnatic blood' as possible (Bharadwaj, 2003). The question arises: Why is the India state inserting the market, and contractual commercial transactions, into a 'domestic moral economy' of exchange and kin relations? These transactions are not mere examples of intra-familial sharing, gift giving, and altruistic exchange, but rather a context-sensitive and contingent way of making sense of reproductive disruption and family-forming strategies. In many instances, intra-familial arrangements are modern ways of doing 'tradition' (Bharadwaj, 2006a), negotiating with patriarchal, gendered and religious injunctions, while conforming to the 'pro-natalist imperative' (Bharadwaj, 2005b). For example, the ancient practice of levirate is still re-enacted in IVF clinics, where the sperm of a male agnatic kin is used to induce conception. The practice of adoption, in the majority of cases, is similarly confined to the wider familial unit for several deeply held cultural reasons (Bharadawaj, 2003).

The neo-liberal state in India, however, is seeking to outlaw these practices for reasons far more complex than mere social reform and the protection of women from the excessive demands made by the patriarchal ideological order. The new laws hold the potential to create a body of medicalized childless citizens that can be both harnessed and garnered for extracting embryos/gametes to fuel the burgeoning global moral economy in stem cell creation. As we shall see later in the book, by putting in place strict informed consent procedures, instituting a further National Ethics Committee, and ensuring the provenance of any potential stem cell lines accruing from human embryos, the Indian state is seeking to isolate ethical sources for procuring raw materials. These ethical sources are imagined as fully informed, rational and autonomous consumers, seemingly liberated by the market from the fetters of traditional and outmoded reliance on familial support for 'assisting life'. These autonomous citizens can now be enjoined to sacrifice their biogenetic spare embryos with an encrypted provenance in the service of a booming 'neo-India' (Bharadwaj, 2006c). The resulting stem lines from 'sacrificed', as opposed to 'killed', embryos forge a global moral alliance alongside the act of self-sacrifice by an informed, autonomous donor. The future of embryonic procurement, stem cell production, and global consumption seems somewhat assured.

Citizens' choice?

What kind of citizen is being constructed? Interviews and prolonged inter-actions with treatment-seekers in the IVF laboratories of the Army and Research Referral Hospital, as well as in an upmarket private clinic in South Delhi, shed light on some possible answers. These interactions provide an insight into the subjective location of individuals and couples in search of assisted conception, and into how they might respond when doubly implicated in the project of 'assisting life'.

Rohit and his wife Jyoti were first encountered on the morning of 29 June 2004. They both hailed from a rural area in the north Indian state of Uttar Pradesh. Rohit, a bright young man of 28, was a 'mere foot soldier', as he put it, in the Indian army. At the time of the meeting, he was posted in the western state of Gujarat but visiting Delhi's military hospital on an expe-dited sick leave for an urgent IVF procedure. Throughout the course of everyday interaction, and one tape-recorded interview, his young wife remained silent, and despite repeated questions, she would only smile and veil her head. With sound cultural reasoning (not to mention a strict patri-archal upbringing), she would not speak to a strange man, and most importantly would not speak at all in the presence of her husband. The couple had undergone three unsuccessful intra-uterine inseminations (IUI), and were awaiting an embryo transfer on the day of the interview. '[I]t could be my shortcoming', he said, 'it is also *vidhi* [destiny], up to God ... it is something technical [referring to the IVF] so you can't predict.' When asked how infertility impacted upon their daily lives, Rohit got philosophical (and he often did so on a daily basis):

> Don't make any weakness so great that it destroys you ... maybe it's God, maybe its *karma* from another life ... clearly society, villagers, don't think ... they will say get married again, keep her and bring a new wife as well!
>
> (Interview material)

Jyoti's glass bangles clinked as she readjusted her veil, which seemed to act as a safety net in a patriarchal environment that demanded unquestioned submission, as well as producing customary and deferential distance between the masculine and the feminine, the auspicious and the inauspicious, the fertile and the 'barren' (Patel, 1994). Rohit continued:

> In the end, the decision is ours, mother, father, it is sad but they can only put pressure on you, it's small thinking, everyone says if you have no children what is this life? You are working for what? They think like that and, who knows, maybe their thinking is tainted with malice and malicious intent.
>
> (Interview material)

Rohit was steadfast in his support for his wife and did ward off tremendous social and familial pressure, but countless months serving away on a 'non-family station' in far away Gujarat left Jyoti alone to bear the full brunt of any social opposition to their use of IVF. Rohit's and Jyoti's predicament did leave one question unanswered: How might this couple, or many others like them, become 'fit' candidate donors, or be turned into consenting individuals that willingly part with their embryos for stem cell research? Rohit was extraordinary in one respect, in that he displayed a free and autonomous attitude in standing up to the societal pressure to remarry. However, when probed on the issue of stem cell research, and some of the moral debates surrounding the potentiality of embryos, the philosopher-soldier resurfaced. 'In Indian culture, killing a creature (*jeev hatya*) is a sin. But if a seed is crushed it is not the same as slaying the tree, that is killing the embryo (*bhruan*).' The statement was ironic, since, as a soldier of the Indian army, he was trained to kill. His subsequent statement was particularly instructive:

> [I]f something from our body can heal and benefit, then it is truly good, it is very good, why turn that into politics? So if you can save someone using my body, it is very good, it is not killing! Indian culture says if my body can benefit you, I can sacrifice my body. If we can save someone through the sacrifice of donating an embryo (*bhruna daan*), then it is good and imbued with a lot of religious merit.
>
> (Interview material)

His training as a soldier, including his professional remit to protect the nation by making the supreme sacrifice, melded with his philosophical musing on *bhruna daan*. His statement unwittingly echoed the underlying ideological basis to an emergent global moral economy that is busy developing legislations, protocols, and ethical injunctions that encode embryos with a moral provenance, which, in turn, could be seen as an expression of informed choice and autonomy. For this reason alone, an iconic moment like the 2001 US presidential decree becomes a point of transformation, as shown earlier, where moral distinctions between sacrifice and killing could, in principle, become the ideological basis for the bio-ethical discourse. However, Rohit's action in seeking to perform a 'sacrifice' is far removed from the principle of an informed autonomous choice to give. Rather, it is an expression of duty, and an extension of his professional life. The duality of embryo and parental body is collapsed, since potential offspring in India, at least in certain contexts, are seen as an extension of the bodily and spiritual self. The notion of sacrifice, therefore, is one rooted in the self and cannot be recast to mean killing of an 'other'. The idea of sacrifice in India, both as bodily and spiritual practice, is ingrained in everyday cultural idioms, conventions, kinship norms, familial obligations and other social practices. Intimate connections between the body and sacrifice are continually reemphasized through story telling, resurrected myths, and, more recently, through mass media and popular culture (Cohen, 2001; Uberoi, 2006). It is possible to conceptualize Rohit's sacrificial gesture as altruism, situated ethics, gift giving, and morally sensitive donation, but these explanations do not come near enough to the core of what lies beneath a willingness to give in Indian IVF clinics.

Excerpts from an interview with an army wife help to identify and clarify the driving gesture that frames a willingness to give. Shanta was having explained to her the difficulties in sourcing human embryos for research in many parts of the world because of the moral and ethical complexities associated with the embryonic form. She interrupted mid-sentence and retorted:

> This will never happen in India! Infertility couples in India will not object [*sic*], people here are very soft-hearted, they will never hurt anyone, infertility couples will not be able to say no, people have been through so much. When you face problems, we Indians face a lot of problems with infertility, they'll accept it, there is no question they won't.

> (Interview material)

Shanta had put her finger on the pulse. It was not altruism that underscored almost all expressions of 'giving' in the IVF clinics, but empathy, coupled with reproductive biographies suffused with a socially disabling search for conception. These were not informed consumers, or rational producers, of embryonic entities, who were consenting to altruistic giving. On the contrary, these were emotional, empathetic, sacrificial expressions, significantly different from the kind of rational sacrifice that the neo-liberal state of exception demands. In all 40 interviews, individuals and couples could not distinguish between clinical research and its application. Despite repeated conversations and explanations describing the nature of embryonic clinical research, their predominant refrain remained that their embryos might 'benefit someone somewhere', 'help alleviate suffering', or, 'take away the pain'. They viewed their embryonic contributions through the register of suffering and healing. The most telling expression of this emerged in an interview with another army wife, Savita, who broke down saying the words 'I want a child no matter what.' A little later, and slightly more composed, she responded to the question of giving up of embryos for research, 'Imagine if I wanted a donor and no one gave, how would I feel? So you should give, there is no problem in sharing embryos … .' Like Shanta, Savita's imagination of what giving might accomplish was filtered through her own subjective experience of suffering the 'pains of infertility', and by empathetically placing herself in the predicament of an unknown other who might, like herself, be forced to lead an 'incomplete life'.

Encounters with middle-class consumers of new reproductive technologies in the private clinic were no different. Rekha, a representative example, is suffering from secondary infertility since the premature birth of her son. She felt that her son, now a thriving 8-year-old schoolboy, would benefit enormously if he had a sibling. Very soon after his birth, Rekha got pregnant again, but she and her husband opted for a medical termination as a result of the overwhelming task of looking after their premature baby. All subsequent attempts to get pregnant, however, had failed. Eight years on, she was riddled with guilt, and even contemplated her secondary infertility as divine retribution for 'killing my unborn child'. Responding to the question of embryonic research she asserted:

> [I]f you're doing it for a cause, research, which will eventually benefit mankind, someone will benefit from it [but] embryos for research should only be contemplated by people who really want to give but then again you can't tell, it's a thin dividing line.

> (Interview material)

Further into the interview, her biography began to surface:

> [After a long pause] [I]n the interest of science, it's okay, after all, some sacrifices were made by some people for research and the advances I benefit from right now, some sacrifices were made … going through this process you think differently, if I had naturally conceived a couple of children, I would not think about it but I feel I am reaping the benefits of someone's sacrifices, you realize the value of something … I now view [the] embryo as a child, if I would have thought like this before I would not have gone for MTP [medical termination of pregnancy]. So going through this difficulty [secondary infertility], you feel differently, till I hadn't gone through this process I viewed it differently but when it actually hits you, you realize how much of a difference it makes to your life, otherwise it's drawing room talk. Only when it hurts, you realize how much it hurts.

> (Interview material)

Rekha's pursuit of conception was riddled with guilt and frustration. She enunciated the moral distinction between killing and sacrifice by splicing it with her own emotional and financial struggle to achieve pregnancy. The MTP stood out as a thorny subtext in her reproductive biography, a decision that she regretted, but more importantly a decision that shaped an empathetic connection with her 'killed' embryo, seen now, in the embryonic form, as a potential sibling for her 8-year-old son. She used the word 'chilling' to describe the thought of giving up the embryo for research, but, equally, felt that sacrifices had to be made. Throughout the interview, she appeared to oscillate between the old Rekha who had opted for the MTP, and the Rekha shaped by the pain, hurt and regret. The 'sacrifice' of embryonic life for the greater good was almost expiatory in her case, as it helped process the guilt of 'killing' an embryo that could have completed her perfect family some years ago.

Consent, contract and choice

Lock explains how the procurement of human materials to make immortalized cell lines was fostered by two international agreements – the Convention on Biological Diversity, and the World Trade Organization TRIPs (Trade Related Aspects of Intellectual Rights) agreement. These international agreements were made legally binding in 1993 and 1994, and led to the globalization of intellectual property laws. Lock further contends that this meant that 'individuals who donate their own body parts for research purposes do not retain property rights over such materials, nor can they participate in any profit that might result from manipulation of these materials' (Lock, 2001: 65). Notable exceptions, however, existed. Lock describes a case where the US government made a patent claim on a cell line created using blood taken from a 26-year-old Guaymi woman (the Guaymi are natives of a remote corner of Panama) suffering from leukaemia. Similar patent claims were made on cell lines obtained from the Hagahai of New Guinea and from native Solomon Islanders. The Hagahai reportedly agreed to the blood donation, and subsequent creation of cell lines and their patenting, on the condition that individuals claiming Hagahai ethnicity would share half of any resulting profits from a vaccine or any other by-product. However, all patent claims on Guaymi, Hagahai and Solomon Islanders were eventually dropped as the resultant cell lines were deemed unprofitable for pharmaceutical business (ibid.: 63–91).

The above cases provide a useful point of departure from which to better contextualize the emergent *biopolitics* of embryonic stem cells in India, in which infertile citizens and their biogenetic capital are being invested with the promise of future returns. On the issue of commercialization and patent issues, the ICMR Guidelines on stem cell research declare:

> Research on stem cells/lines and their applications may have considerable value. Appropriate IPR protection may be considered on the merits

of each case. If the IPR is commercially exploited, a proportion of benefits shall be ploughed in to the community which has directly or indirectly contributed to the IPR. Community includes all potential beneficiaries such as patient groups, research groups, etc.

Unlike the US government, and, indeed, the indigenous populations who collectively bargained for downstream IPR benefits in the cases above, the Indian state's decision to pre-empt any such moves might be seen as far-sighted. However, there is a morality of another kind at work in the economic calculations of the Indian state, namely, how might the informed citizens be enrolled in the production of knowledge which is both promissory and profitable? The projected benefits are twofold. First, they offer the tantalizing future promise and hope of cures for a spectrum of disorders, and, in that respect, the renunciation of a spare embryo is a worthwhile investment. Second, the promise of future returns is made tangible, not merely as a therapeutic breakthrough, but also by building commercial and economic stakes into the gift itself. The contract with the state, through the modality of consent, is predicated on the creation of a citizen imbued with the will to choose morality and profitability, as opposed to sacrifice in the hope of alleviating suffering. Nevertheless, given the official interest in IPR and profit sharing, the Indian state appears to be working at odds with the global drives to extract commercial potential from embryonic entities.

Like the Guaymi, Hagahai and Solomon Island patents, it would not be at all surprising if, in making the transition from knowledge to practice, and from science to commerce, the IPR issue is dropped altogether as logistically, commercially and economically untenable. This becomes more probable as private capital, as opposed to state investment, flows into the biotechnologies. In addition, India is increasingly being seen as an embryo-surplus nation (Jayaraman, 2001). For example, it is argued that there is a rich and steady supply of spare human embryos emerging from the many state-of-the-art IVF clinics in the country. IVF in India is almost exclusively a private sector enterprise, and any moves to make IPR and profit-sharing a necessity is bound to attract severe opposition. The private sector stem cell initiatives are, on the contrary, adopting innovative solutions, whether as a matter of conscious strategy or serendipitous corollary, that circumvent the IPR and profit sharing initiatives in the proposed guidelines. This becomes particularly clear on closer examination of the exact modality of procuring embryos for research, the process and nature of informed consent obtained, and how the views of the women and the men who sacrifice their biogenetic capital are framed in clinical encounters.

Through the course of fieldwork in a privately owned stem cell research facility in Mumbai, it became clear that spare embryos were sourced from within the 'in-house IVF practice' that functioned as an extension to the larger stem cell research programme. The admission by the clinician in question (Dr Anand) was candid. He saw nothing remotely problematic about such practice. 'After all', he contended, 'there was a well-informed process of informed consent in place and besides the infertile treatment see-kers were more interested in a quick pregnancy than in larger philosophical issues surrounding potentiality of human embryos.' This was to be later corroborated by another fertility expert in Mumbai (Dr Bharat). When probed further for clarification, Dr Anand relented and admitted that the notion of informed consent, so-called standard clinical practice, can be ren-dered hugely problematic, as in the final analysis it was down to the indivi-dual clinician to determine how the facts were explained. In the next breath, he followed this contention with a dramatic admission:

> I can sit here all day explaining to a couple how good it is to donate embryos for research or equally convince them otherwise if they at first instance appear to be hesitant or unsure and give me the slightest impression they shouldn't even think about donating. Either way it's up to me how I play it, I can convince a person either way.
>
> (Interview material)

In one swift move, he laid bare the contours of the informed consent process and its constructed nature, and the discussion returns to this later in Chapter 6.

Dr Anand's colleague and one of the senior scientists in the stem cell team, Dr Richa, similarly observed that the informed consent process could be problematic, but that, if counselled properly, the donors see nothing wrong in giving up 'spare embryos' for research:

> [M]ost of the patients are happy about donating it, especially when we explain to them the concept of embryonic stem cells and how, if the centre takes off, and it really starts working as we are expecting it to work, then it really would be the future of medicine and it would really give a lot of cures for a lot of diseases. So, yes, most of the patients do agree that otherwise [we] discard the embryos. So rather than discarding it, if you can use it for something better that can be useful in the future, then why not? So most of the patients are receptive. You do get a pro-portion of patients who do not agree to donate. Especially like giving their embryo once it has been formed to some other couple. Yes, there is resistance because they feel particularly that is their child. But if they have been counselled about donating their embryo right from the beginning or donating their oocyte, we do have an oocyte-sharing pro-gramme. So if they have been counselled right from the beginning, then they are open about it. But rather than giving their embryo to another woman, they are still, I feel, more open to donating it for research because they feel that baby would be born like if they had given it to another woman. And then – so there are some ethical issues they have in

mind that, okay, what rights do we have to that baby? Would that be our baby, and things like that. So they are more comfortable, I think, donating it to research in those cases. But if they have been counselled for embryo donation or egg donation before the onset of cycle, then they are quite okay about it.

(Interview material)

The crux of the issue lies in counselling and preparing the donor to accept the renunciation of embryos as something good and worthy. However, a longer conversation with Dr Anand exposed specific details that revealed that a lot more than the mere giving of information and eliciting of consent was at stake:

DR A: I think this holds true even in the Western countries. I mean, although we say things like informed consent, ultimately I think you – you [the clinician] are a part of the decision-making.

AB: Hmm, true, yes.

DR A: It happens.

AB: Yeah.

DR A: Even – even though you want to be completely out [of the consent giving process], it's not possible to be completely out, because the couples often ask you, 'Well, what would you do?'

AB: Yes.

DR A: Letting them know it's immaterial what I do because I'm not in a sim – similar situation.

AB: Yes.

DR A: Somebody who would suggest, it's completely out of the picture. I don't know because I cannot put my mindset into yours. I cannot think because what you are thinking now is in that frame of mind. I cannot put – I can only tell you the pros and cons, explaining the picture you know, what it is.

And I – I choose the couple to whom I talk. Even if I have a couple whom I think don't come from that background, education-wise, to understand the stem cells, I don't talk about it. I freeze it today. Because what happens is maybe they didn't look at me with a true understanding, they – is he going to use them in any case if I say yes or no? How do I guarantee this process?

Because they are the ones, who – you know, or some couples are so stressed, so stressed, I don't talk to them either. Because they're so stressed that in case of that treatment fails, then has he [the clinician] used the good embryos for them ... or giving me the bad embryos? Or you know, has he compromised my treatment? So I talk to couples whom I think would understand this, and they should have the minimum – (*sic*).

AB: Sure.

DR A: Qualification to enter into my thing to talk to them otherwise I just freeze it for them.

AB: Right.

DR A: I don't expose them to the embryonic ten per cent because – what happens, they may judge me. They may think that, you know, he's – is he using me as a guinea pig, trying to get eggs out of me and will only put two in me and take all twenty?

It doesn't happen. It doesn't happen.

AB: No.

DR A: But at the same time I am putting my credentials at stake and either the very people, the treatment fail, well, the one in ten people, you know, who talked about something, maybe embryonic stem cell and they use my embryos for this. So it makes good headlines in the media.

'And I went through IVF and they used my embryos for stem cells, that's it.'

And I belong to a company which has got a very high re – reputation and stakes in the market –.

You know, one headline and –

Share markets.

It – it affects their share price.

AB: Yeah.

DR A: So, I have to be very, very careful what I talk because either the very people – they may – see there are lots of people just waiting to gun down the company because we've got bio and life – life sciences and bio-technology (*sic*).

(Interview material)

Many treatment seekers in this research echoed Dr Anand's concern with the use of embryos for inducing conception and for stem cell extraction. The responses that emerged remained focused on the contention that, 'so long as our treatment is not compromised, we don't mind giving away whatever they need'. Dr Anand repeatedly reiterated how crucial it was to constantly reassure the donors that only the best embryos were used for their treatment and the rest, the 'less than perfect' embryos, were set aside for stem cell extraction. In response to the proposed ICMR ban on generation of embryos for research, Dr Richa commented how the quality of embryos affected the quality of derived cells, and that a method ought to be conceived that allowed access to the better quality embryos that were routinely used up by the IVF cycles:

It is difficult because there is a high percentage of population that is still illiterate. So trying to explain to them what this would do in future, it is difficult. Especially because it's difficult to explain to them about the disease process, then at what stage we are taking the embryo, and those that would not affect their treatment. It is something that is in surplus after it has – like, the best quality embryos would be transferred to them because their concern is, would you take only the best embryos for your research and transfer the poor embryos into the uterus? So they need to be counselled thoroughly. And we make sure that we transfer the best embryos and only the poorer embryos ... are left behind.

(Interview material)

The research could not ascertain what quality controls and methods were put in place at this facility to make the embryo selection process transparent. Nevertheless, what is relatively clear from the foregoing is that informed consent is much more than just good medical practice in the domain of *biocommerce*. As the long excerpt from Dr Anand's interview reveals, moral and commercial concerns underpin volatile and market-driven corporate reputations. Such reputations can be undone on the back of one unfavourable response. The morality of embryonic procurement, therefore, has to strike the delicate balance between keeping the 'gifting' donor informed while ensuring that adequate moral and ethical concerns are factored in as 'costs' in order to maintain market leadership, profitability and commercial viability. If ethics and morality are necessary business costs, then an informed and happy donor equals good business sense. However, there is another side to the corporate investment in biological sciences that is both result-oriented and profit-driven. When asked what were the pressures and demands of researching a new and cutting-edge science within a privately-funded corporate set-up, Dr Richa explained:

They demand results. They demand results, absolutely. It is because any research takes time. And it's like the experiment that you start with would give you positive results. It will be 100 times you will get a negative result. I will give you an example. I was working on one thing and I said I can write a thesis on how not to do it. What are the things that would not work, I could write a thesis on that. So – but being a corporate set-up, they are used to products and manufacturing. And they want things to start making money immediately, as soon as yesterday. And this being a new science, yes, everything takes a long time. Each culture runs for days together. Sometimes three months, four months. So until that time we are not able to give them the answer whether that stem cell line is growing, not growing, what's happening to it because it has to go through several passages before we can say that it has really maintained its 'stemness'. A lot of stem cell lines, they would start growing properly. But then, on the way, the cells may start differentiating ... and again the company wants to be first in giving any product out. So, yes, that pressure definitely. But then once we are giving you so many facilities, they do expect something. And nothing comes free of cost.

(Interview material)

Dr Bharat, in Mumbai, went a step further, and exposed the less than salubrious under-belly of the so-called informed and consented 'donation' of embryos in the private sector research facilities. He began by asserting the usual stories about the profound social stigma and ostracism faced by infertile couples, and how the bulk of the pressure remains on women regardless of the exact reason for reproductive disruption. 'If a woman with blocked tubes comes to you,' he continued 'you can only do an IVF for her, there is no other option.' The cheapest IVF treatment in India costs about £1000 per cycle. In his estimation, more than 50 per cent of infertile middle-class patients struggle to pay this price tag. He then proceeded to recount the number of patients he sees in his clinic, asking for help and threatening suicide. The social stigma and ostracism associated with infertility in India are well documented (Riesman, 2000; Bharadwaj, 2001) and if it were not for this research, Dr Bharat's assertions could easily be criticized as sensationalist and exaggerated. However, it was his subsequent assertion that unravelled the problematic aspects of informed consent in Indian fertility clinics in the age of stem cell research. He continued:

[I]f I go and tell 1000 infertile patient who need IVF and can't otherwise afford it that I'll give them free IVF, they'll say yes. If I tell them half of their embryos I am going to use for stem cells, do you think they'll say no? They will jump at the chance and sign any piece of paper. Somebody comes to me and says I need materials for stem cell research and I will fund IVF cycles in return, what do you think I will do, say no? Here patients are prepared to sign blank forms, forget informed consent.

(Interview material)

Dr Bharat was espousing an ethic that saw no harm in using embryos for research so long that everyone involved benefited. For him there were only winners in this game and no losers. Dr Richa, in fact, went on to suggest that, in the desperation to get pregnant, patients seldom want to spend time learning more about the consent process:

[H]alf the time when we are explaining to them, they don't want to listen to us. They are, like, tell us where we need to sign. But to be on the safer side we always want to explain to them because tomorrow they shouldn't come back to us and say, doctor, you didn't tell us this. So, yes, the scenario's very different from the West, where the patient would have read on the internet the thousand things possible and then come to you with reading questions. Your – it's a very small population of patients that would come to us with reading questions that – doctor, we read about this thing on the net. Do you offer this technique or how would that help us or what are the advantages and disadvantages? Most of the patients would just agree to what you tell them. And you have to tell them to think about it, sit down, think about it. If you have any questions, do come back and ask us.

(Interview material)

A Delhi-based fertility and gynaecology expert, Dr Shah, similarly commented that in India:

DR S: Hmm, English inform consent, nobody can understand, because in India we sign papers without reading and it's the done thing ... There's no Hindi equivalent or no other language equivalent, so the informed consent [is] a joke, it's a very dirty joke that the profession plays on the individual, and they couldn't be bothered if you have the babies, it doesn't matter what you do with the rest ... So they're [fertility treatment seekers] extremely exploitable, so if anybody was interested in the commerce of stem cells and this, you know the future of stem cells, so anybody who's interested in the commerce of stem cells steps in ...

AB: So you are virtually saying that people, individuals, couples, are not informed?

DR S: Yes, that's right, that's right, so if they want, what they give me, say, 14 embryos in a cycle and I use two, I'm left with 12 embryos, I'm reasonably sure they would, if they were ever going to conceive in the next three or four cycles of embryo transfers. So then that leaves another six embryos, that's eight embryos, what are you gonna do with the remaining ones? Stem cell research! ... you see nobody knows how many embryos they have, nobody knows how many eggs they have, nobody bothers to pick up a discharge summary, nobody knows, there's no cross-checking.

(Interview material)

The mockery of informed consent process appears to be in aid of putting in place a practice in which embryonic stem cell research could easily become parasitic on profound social suffering. Through the course of this project, and in earlier research (Bharadwaj, 2001), scores of infertile women in fertility clinics in India expressed their fear of abandonment, and saw in the technology of IVF the last remaining hope to salvage their unhappy, childless marriage, while strategically managing social ostracism, stigma and social exclusion. Even within the conjugal context, couples struggle with extremes of emotional, physical and financial depredation inflicted by invasive and expensive assisted reproductive technologies. Though purportedly rational consumers of conception technologies, their consent to comply with the treatment regime was almost always a feature of their 'delegitimated' social location. The stigma of infertility in this sense becomes what Kleinman calls '*sociosomatic*', resulting in the '*delegitimation*' of the stigmatized rather than rendering their social identities spoilt (Kleinman, 1995). That is to say that, though stigma – as in the West – remains a moral category, the process of delegitimation affects the person, the family and all aspects of social intercourse.

Conclusion

The conception story of embryonic stem cell biotech in India is embedded in huge potential for human exploitation, and while it may not seem to be as brutalizing as the trade in human organs in India and elsewhere (Cohen, 1999, 2001; Das, 2000), the very creation of the entity, the 'spare embryo', raises some problematic questions. For instance, not every IVF procedure requires the generation of spare embryos, and once created, there is the attendant problem of freezing such entities. Freezing embryos is seldom a preferred option as the quality of freezing programmes can vary greatly and compromise success rates. As one clinician in Delhi pointed out, 'to freeze an embryo is to also freeze its fertile potential'. Despite this, frozen embryos are gradually becoming the norm in fertility centres across India (save in those smaller clinics which cannot afford to have freezing programmes). It is this gradual shift from an embryo to spare embryos that makes the science of embryonic stem cell research morally contentious. Desperate infertile couples are being sold a dream of free IVF cycles (and some may not even be candidates for IVF), so long as they agree to the creation of spare embryos. Their bodies' reproductive potential, or the lack of it, is purchased through a circuitous route of a free-IVF-for-spare-embryos transaction, only to be repackaged in the language of 'informed' and 'consented' gift, of giving and donation. In this context it is worth revisiting the role played by the Indian state and the ICMR draft Guidelines on stem cell research and regulation that expressly forbid the generation of embryos for the sole purpose of obtaining stem cells. This leaves only one potential exploitable source, the infertile IVF treatment seeker, for the generation of spare embryos. Contrary to the promissory note of future benefit sharing offered by the state, the potential biotech futures that these infertile bodies make imaginable remain out of bounds for the 'informed' donors, as they sign away access to future breakthroughs derived from their embryonic contributions in exchange for free IVF. The contract between the consenting citizen and the state is bypassed. Nancy Scheper-Hughes, citing Veena Das, argues that the neo-liberal defence of the individual right to sell could be problematic, since in all contracts there are notable exclusions (Scheper-Hughes, 2000: 197). In this respect, future prospecting remains the sole domain of corporate pharmaceutical players (Brown and Michael, 2003; Hedgecoe and Martin, 2003), who underwrite IVF with one eye on the generation and procurement of spare embryos, and the other on the potential, profitable futures that these spare embryos could create. Whether sacrificing citizens, who are now enrolled by the Indian state as an ancillary to the production of profitable therapeutic futures, will remain a 'notable exclusion in the contract' or whether they will become a 'benign reality', is something that will become apparent as and when these futures come to fruition.

The state of exception invoked as a result of the August 2001 American presidential decree was just that: a state of exception that allowed 'bare life' to be harvested in the pursuit of research and profit. All subsequent global production processes have emerged, or are morally required to originate from, ethically unproblematic sources. In recapitulating bare life into the realm of law, morality, economics and politics, market-driven, neo-liberal formations like India are engaged in engineering informed, autonomous and rational citizens whose bodily surplus can be meaningfully and legally harvested in the pursuit of science, healing and profit. The precise reasons why individuals consent to giving embryonic materials, and how implicated are their own biographies in such purportedly moral transactions, are relatively unaddressed. These emotional, as opposed to rational, consumers of conception technologies, the 'producer owners' of embryonic materials, are empathetically connected to the cause of giving. They see their sacrifice as being for the greater good, for alleviating the suffering that they know and understand all too well. The state, on the other hand, demands sacrifice in the service of a bureaucratic rationality that views both ethics and morality as necessary cogs in the market-driven production process. The morality of these men and women is not linked to the production of guilt-free consumption in the neo-Global order (though they unwittingly contribute to its generation), but, rather, empathetically linked to an 'imagined community' of sufferers. This morality is in fact about being able to see the self in the other (Bharadwaj, 2006a).

The global moral economy of embryonic stem cells can become viable only so long as human embryos are seen as being sacrificed, as opposed to being killed. The difference is not merely semantic, but is ideologically driven in the pursuit of creating global normative benchmarks of ethically acceptable and viable standards. That these standards are often market-driven, with the promise of supplying morally unproblematic commodities that facilitate guilt-free consumption, is the singular strength of the moral economic sensibility. Given the globally variable status of human embryos, the conceptualization of the human embryonic form as the sacrificial citizen mobilizes moral sentiments, nationalistic imaginings, and patriotic fervour, unlike the imagery of embryo as killable, as bare life preceding the human/ divine law. While these moves are not expressly stated, though they can be gleaned from the existing moral panics around the world on the question of embryos, human oocytes and stem cells, they are implicit in the global political economy of embryonic stem cell research and development. The morality invoked in using spare embryos for ESC generation around the globe, and in locales such as India, remains problematic. These spare embryos emerge as tainted entities from the point of conception, as their surrender for research, or adoption by another infertile couple, is inextricably bound up with the helplessness permeating the local moral worlds of the infertile. Their consent is informed by a very different social reality that is never addressed in the clinic, but, nevertheless, is often exploited for extracting spare embryos. The question remains: Can 'spare' IVF embryos in India ever provide a morally sound foundation for the kinds of global biotech future scientists are trying to forge? In a post-Cold War, dis-located, global order, the calculus that drives ethical and legal convergence between nation–states, in search of a morally and ethically unproblematic production and circulation of bio-genetic entities, can no longer be articulated in nationalistic, local or regional terms. The neo-liberal ethic operates on a global scale, where shared morality and ethics become necessary conditions for the international circulation and accumulation of capital. As Sunder Rajan (2006: 83) argues: '[M]any of the tactical and strategic articulations of the Indian state tend not to be "resistance" to global orders of technoscientific capitalism, even while they might rescript hegemonic imaginaries in ways not imagined', or, in the light of the foregoing, 'now becoming imaginable'.

5 'Miraculous stem cells'
The liminal third space and media rhetoric

Introduction

Media representations of techno-scientific advances are important pointers towards understanding how cultures differ in what is regarded as acceptable (Haran *et al.*, 2007). The media provide an important source of information about new medical and scientific research, and often are the arena within which policy battles are fought (Wellcome Trust, 1998). As a result, the media are sometimes the focus of intense lobbying from powerful interested parties. There is, of course, no necessary correlation between public opinion, discourse, or policy, but the media do influence which, and how, events get defined as public issues (Petersen, 2002). The media do not intentionally deceive or mislead their audiences, but necessarily limit what they receive. They frame stories through the selective presentation of one set of themes, oppositions, 'facts', and knowledge-claims as compared to others. Sometimes, while some new issues are highlighted, as with the debate surrounding the cloning of Dolly, or the use of spare human embryos for stem cell research, only the rhetoric changes to reflect and little else. Kitzinger and her colleagues conclude their study of how the bioethical debates surrounding embryonic stem cell research were framed between 2000 and 2005, by observing that:

> [A]lthough a major controversy was aired, many fundamental questions were left unaddressed. The coverage left the existing system of medicine and the scientific enterprise itself largely unchallenged. Although appearing to represent a range of conflict, certain positions, including the position of ambivalence, were largely silenced. Although including quotes from an apparently wide range of 'balanced' sources, some voices were systematically marginalised.
>
> (Kitzinger *et al.*, 2007: 227)

As Petersen notes in his study of news media portrayals of human cloning after Dolly, the prominence of this event ensured wide coverage and debate. However, it tended to focus predominantly on human not animal cloning, and spent little time critically analysing or reassessing the original research. It was preoccupied with the imminence of human cloning. The effect, he concludes, 'was to reinforce a particular public definition of cloning, which may have fuelled fears about the imminence of human cloning' (Petersen, 2002: 87). Some scientists, such as Richard Seed, were depicted in stereotypical terms to conform to a script that had more in common with Frankenstein (Turney, 1998). There was little considered discussion of the wider issues at stake:

> News coverage provided little opportunity for readers to question the portrayal of cloning as duplication, imitation, 'genetic stencilling', and the like, or to reflect in any depth on what is at stake in cloning research and whether it really offers the threats that many commentators suggested.
>
> (Petersen, 2002: 87)

One result of this is a belief in the power of science coupled with an increasing mistrust of the scientists who wield it. There is an assumption that scientists will, if unchecked and unregulated, deliberately set out to clone humans in unethical circumstances – indeed, a powerful trope. Haran (2007) suggests that scientists attempt to establish the boundaries between 'maverick cloners' and mainstream scientists by lobbying the media. For example, Panos Zavos claimed in 2004 to have produced a cloned embryo by fusing skin cells from an infertile man with an egg taken from his wife. A group of prominent scientists, through the Science Media Centre and the Royal Society, appealed directly to journalists to be as sceptical as possible about this claim, particularly with reference to therapeutic cloning which would be illegal under UK legislation. The scientists used their status not only to distance themselves from this 'cowboy cloner', but also to establish the boundary should future mavericks appear on the scene. As Eriksson discovered, British scientists operate with firm boundaries between those 'inside' and those 'outside' the acceptable research community, and feel it 'crucial to avoid giving scientific legitimacy to scientists that they described as "mavericks"' (Eriksson, 2004: 25). The similarities between these events and those surrounding the Pusztai affair in the late 1990s are clear. At that time the Royal Society also stepped in to take control over the rupture to the legitimating fabric of peer review to dismiss claims made via the media by Dr Arpad Pusztai that some GM potatoes contained genes that significantly affected the organs and metabolism of rats (Glasner and Rothman, 2004; Murdock, 2004).

In part, accounts of advances in the new biotechnologies in the media depend upon unelaborated constructions of future promise, be that for good or ill. These are co-constructed by key stakeholders alongside the media, although the weight given to who speaks with an authoritative voice is not pre-determined, but emerges as the debates develop. For example, Professor Hwang was forced to resign from his senior post at the Seoul National University following the discovery that he had falsified at least nine of the 11 embryonic stem cell lines that he had originally announced in his paper in *Science*. He had even announced the world's first cloned dog, an Afghan hound called Snuppy, in *Nature* in August 2005, leading some commentators to crown him the 'cloning king'. However, prior to his downfall, Hwang had been fêted as a national hero in South Korea, and he had managed to convince important stem cell researchers, including those who had peer reviewed his work for publication, that his ground-breaking research was beyond criticism both scientifically and ethically. The media response to the publication of his original papers was to follow the bandwagon initiated by his peers. It left little room for ambivalence and cautious optimism, something also missing in the coverage of the UK Chief Medical Officer's Expert Group (the 'Donaldson Report') in 2000, which resulted in cloning becoming regulated under the Human Fertilisation and Embryology Act (Kitzinger and Williams, 2005). Furthermore, there was very little attention given, unlike in the Hwang affair, to financial prospecting as part of the imagined future. As Haran and her colleagues conclude: '[T]he mass media are increasingly being incorporated into the daily operation of early twenty-first century biosciences' (Haran *et al.*, 2007: 180).

The development of stem cell technology in India, on the other hand, is contained in discursive realms such as nationalist agendas of the Indian state, emerging governance/legal frameworks and the political economy of health. However, a wider social response to these developments is yet to emerge. This notable absence of public debate about religious or theological concerns with the manipulation of biogenetic entities like human embryos stands in dramatic contrast to the moral trepidation surrounding embryonic stem cells in countries such as the United States.

In India, the adulatory media renditions on embryonic stem cells (ESC) have surprisingly (re)produced the narrative templates that generated much hype around the proliferation of New Reproductive Technologies (NRTs) (Bharadwaj, 2000). Very often, deep-rooted political and economic forces shape media outpourings on science and technology reporting in India (Bharadwaj, 2000, 2002), and there is a notable absence of ethical or moral debates about the technical manipulation of human biogenetic materials. In April 2002, Dr Firuza Parikh, of Reliance Life Sciences, was quoted by PBS, an American private, non-profit media enterprise, on its *News Hour* programme as saying 'as yet, stem cell work in India is apolitical' (Lazaro, 2002). The programme reporter Fred de Sam Lazaro of the Twin Cities Public Television reiterated that, 'unlike their American colleagues, Indian scientists are not caught in a societal debate over when life begins' (ibid.). Similarly, Bala Subramanian, a cellular biologist who works on adult corneal stem cells at an eye hospital in the south Indian city of Hyderabad, was quoted in a *Washington Post* article as saying:

> [O]ur society is liberal in areas of scientific work. We will not face any opposition ... India already has the expertise to work on stem cells, and

fortunately there is no bar on the use of embryos and aborted fetuses [*sic*].
Nothing stops us now from becoming a leader in this field of research.

(*Washington Post*, 2001)

In the same vein, Firuza Parikh was quoted again, this time in the *Washington Post*, asserting, 'We are for embryonic stem cell research. Religious, cultural and political circumstances here [in India] are not in conflict with our work.' Unlike in the United States and the United Kingdom, where increasingly ethical concerns are being 'built into' new life forms such as stem cells as a matter of both innovation and market strategy (Franklin, 2003b), Indian scientists and the media – unencumbered by the lack of public debate on the subject – are consistently promoting an apolitical, liberal image of scientific work. However, as discussed earlier, global ethical concerns are being built into the ESC life forms in India, and these might be gainfully conceptualized as a pragmatic response to the emerging moral economic reasoning around the globe, as opposed to a cultural response to the question of life and its beginning.

A few notable differences between media reporting of new reproductive technologies (NRTs) and embryonic stem cells (ESCs) have emerged in the recent past. Unlike the adulatory media reporting of the thriving clinical application of NRTs in India, the journalistic accounts of embryonic stem cells have, in large part, remained focused on issues of regulation and ongoing laboratory-based research in various Indian cities. However, this silence was broken when results from Dr Geeta Shroff's clinical trials began to trickle into the local and global media. An unprecedented furore ensued, as noted earlier, that ranged from utter disbelief, to casting Shroff and her scientific work as maverick, unethical, and dangerous.

This chapter captures a vignette of the ongoing media reporting on ESC research in India and compares it to some complementary developments in the United Kingdom. In so doing, the chapter pays particular attention to the 'third space' of stem cell science and the disparate networks that contain and sustain it. This is an anomalous space that goes beyond a simple local and global distinction. It is a space policed by desire for and against liberal rules and governance. A space that accommodates conflicting demands in a globalized research system. The chapter argues for the presence of a 'liminal temporality' as an essential component of this emerging, biotechnologically driven, third space. It is argued that these scientific third spaces owe their existence, in part, to liminal time that enables the anti-structural nature of scientific work. The network of actors/objects/discourses that make up the third space of ESC science in India (and perhaps elsewhere) is enveloped in a 'between-time' that shapes the anti-structural nature of this spatial configuration.

The *liminal third*: revisiting anti-structure

A third space can be an opportunity that did not previously exist. It is sheer potential. It is locatable at a confluence, a site of convergence, with multiple intersecting pathways. Third spaces are creative, generative, and, most of all, unpredictable. An ethnographic excursion, or an anthropological contention, lived multiple identities, a new and daring idea, or opportunities to challenge, engage, criticize and disrupt, are different ways of creating and inhabiting third spaces. Third space is a departure from a rigid script; it is an improvisation. The permutations and combinations are limitless, and so are the consequences and outcomes. Edward Soja sums up the third space as a 'limitless composition of lifeworlds that are radically open and openly radicalizable' (1996: 70). Third spaces are not an in-between position for Soja, but rather they reflect a move away from either/or to both/and/also. Third space is an open-ended space, and it is transformative in that it seizes the normatively opposing, and rebirths them as new in an ever evolving third spatiality. The anthropology of reproduction has argued for some time now how the very act of reproducing introduces a 'difference' (Strathern, 1992; Franklin, 1997). This 'reproduced difference', a state of 'both and also', it may be argued, is the third element that destabilizes the binary and engenders a third space.

What would a third space look like when related to the state of anti-structure and accompanying liminal temporality? For Victor Turner (1967, 1969, 1974, 1975), anti-structure was a state of liberation of human capacity for cognition, affect, volition, creativity. It was liberation from normative constraints. In such schema, prophets, artists, and to these we can now add experimental science and scientists, are marginal people existing in a liminal temporality. Turner saw them as quintessential 'edgemen'. Anthropology has long understood that a between time, liminal time, is fraught with danger and opportunity (cf. Van Gennep, 1960). This temporal dimension opens clearings and spaces that can transgress the structure, the normative and even ideologies. Such an anti-structural realm, fraught with danger and opportunity, has the potential to create, ignite, consume and transform social spaces into third spaces. Turner likened such anti-structural opportunities to an institutional capsule that contained the seeds of future cultural developments and social changes.

However, the realm of anti-structure is not devoid of all structural constrains. For example, as Bakhtin (1984) argued, 'during the carnival time life is subject only to its laws, that is, the laws of its own freedom'. Carnival is liberation from the norm of etiquette and decency imposed at other times; it is liberation from the official structure. In a carnival, the opportunity and danger of market language, and grotesque imagery, is pitted against (are anti) the normatively decent, and subject to a structure of the subaltern – the underdog as opposed to the official. Turner (1967) clarified the presence of a very special kind of social structure between the instructors and neophytes, one that required complete authority and complete submission. In this respect, states of anti-structure and that of liminal temporality are not too dissimilar from third spaces, in that, while radically departing from the rigid, the predefined, and most importantly the opposed, they nevertheless operate under unofficial 'third' structured conditions. If, by the liminal anti-structural, we understand a departure from the official structure/norm, then there is good reason to reconceptualize third spaces as temporally liminal, where rules are constantly critiqued, rewritten and articulated. Unlike the liminal, where there is an eventual return to/reincorporation into/restoration of the structure ending a period of separation, a third space remains liminally open-ended. In arguing for the *liminal third*, therefore, the chapter acknowledges a space that is dangerous, in that it is unpredictable and an open-ended opportunity not only to appreciate but also to critique, challenge and change.

Encounters in the *liminal third*: embryonic stem cell research in India

The relationship between a future temporality and the ongoing scientific work on ESCs in Indian clinics and laboratories is a recent example of liminal third spaces. It is in these spaces that the projective imagination of biotech futures is made tangible through the clinical application of embryonic stem cell entities. In many respects, the science of stem cells – at least in certain quarters in India – inhabits futures that Euro-American countries are yet to colonize, or to politically and ethically regulate. The chapter will attempt to show how these anomalous, unruly, liminal third spaces challenge the very idea of what is to count as local, global and good science. In so doing, the chapter examines how various stakeholders, from scientists, through journalists, to regulatory agencies, seek to radically renegotiate the ethics of scientific engagement.

Miracles in the dark

On 3 April 2005, the Sunday edition of the prominent English language daily, *The Indian Express*, carried the following banner headline: 'Stem Cells: "Miracles" in the Dark'. The piece reported claims of successful results in 24 cases through the application of embryonic stem cells from Dr Geeta Shroff's south Delhi clinic. The article also reported that the Indian Council of Medical Research was beginning to ask questions, and was eager to check the veracity of these claims. The use of the words 'miracles' and 'dark' in the headline is significant. The news report, casting aspersions on the scientific truth claims, cautioned that in absence of any concrete guidelines, gullible patients could end up as guinea pigs. To bolster this claim, the news report compares this 'small-time clinic in Delhi claiming to treat dangerous diseases to getting patients back on their feet' with the All India Institute of Medical Sciences (AIIMS), a respected public teaching hospital in New Delhi, where, in the recent past, autologous cell transfer in cardiac patients was achieved (in which cells are derived from the patient's own body and injected back during cardiac surgery). The ICMR had earlier distanced itself from critical reports in certain sections of the media by claiming that it had no 'knowledge about the AIIMS research'. However, the subsequent media clarifications and statements were unequivocally supportive of the AIIMS' breakthrough, and this is further touched upon in the next chapter.

The use of autologous stem cells in cardiac surgery was viewed as ethically unproblematic, as compared to the embryonic entities that 'the small-time south Delhi clinic' was seen to be using. The emphasis on the clinic being 'small' remained the subtext that compromised the quality of the scientific breakthrough reportedly being achieved there. The news report depicted Shroff, and her clinic, as maverick and too good to be true. Two further news items appeared alongside this news story. The first featured claims by a private hospital in the suburban outskirts of Delhi, that three years ago it used autologous stem cells derived from patients' bone marrow during heart by-pass surgery and had since published the findings in the *Indian Heart Journal*. The stem cell transfer proved successful, and there was significant improvement in the areas rendered dead by the heart attack in all ten patients.

The second news item focused on a breakthrough made by an individual surgeon based in a Delhi public hospital. He was reported as 'quietly repairing and regenerating organs with the help of stem cells without any funds or government support'. The doctor was granted a US patent in 2001, after decades of work had remained unrecognized at 'home'. Earlier, in 1999, his work was published in the *World Journal of Surgery*. The report claimed that a similar paper had been rejected by the same journal in 1990 as stem cell use was not then a documented procedure. The doctor is quoted as saying 'I had to take the stem cells out of the article for getting it published' (*sic*).[1] While all three scientists/clinicians were working in an, as yet, undefined, unregulated and grey area of science, largely unacknowledged, and were not validated by the state and its various legislative organs, only the first clinician, Dr Shroff, is singled out as a 'maverick' in the news report. Her research was not peer reviewed, and could not be considered ethical since it did not involve the use of adult stem cells. Despite Dr Shroff's claims that the trials involved working with terminally ill patients, and that the technique showed good results in such cases, both the newspaper reporter and the ICMR remain unconvinced. The science in this case remains firmly entrenched in the third space from where the clinician articulates the invitation: 'tell us what is wrong and we won't do it' (*Indian Express*, 2005). With no response to this request forthcoming from the ICMR, Dr Shroff persists in face of official opposition.

The bone of contention, however, lies in the use of embryos for research, and in the fact that the science, and the technique itself, have not been peer reviewed. The question of using cells from embryos is especially interesting, as there is, as yet, no clear ethical regulation prohibiting or restricting their use. On this issue, the state, and its regulatory agencies such as the ICMR, are ironically drawn into the same liminal third space as the embryos and their frozen futures. The rapid proliferation of stem cell technologies in India has left a gaping absence of structure, as the existing ethico-legal injunctions fail to adequately contain and regulate research and application. The guidelines that are in place, as shown later, either do not fully regulate the proliferation of ESC technologies, or are creatively reinterpreted by the scientists and clinicians. The patients, in between illness and health, relatively unprotected by law, are similarly citizens of a third space. They are subject to its internal structure and future promise of an eventual cure. The clinicians, who, despite their marginal, maverick and 'edgeman' status, continue to work. They are unrecognized but unencumbered and benefit from the *liminal third* location.

Row over miracle cures

On 18 November 2005, the *Guardian* newspaper reported the international 'row over doctor's miracle cures' using embryonic stem cells in India. Dr Shroff's stem cell therapy became an overnight global curiosity, with newspapers as far away as Australia and the United States reporting on her stem cell treatments. The *Guardian* reported that the 'West urges curb on Indian clinic's untested treatment' while the 'controversial stem cell work gets patient backing'. The 'Western' critics in the report described the use of embryonic stem cell materials as 'highly implausible', and the success claims as 'extravagant claims'. The patient testimonies were explained away as a cynical ploy on the doctor's part to exploit desperate and vulnerable patients by offering them 'false hope'. The construction of 'desperateness', through media accounts, in subjects under the biomedical dispensation is not new (Franklin, 1990). The familiar trope of the 'desperate dupe' was resurrected to great effect in the report. The *Guardian* correspondent, Randeep Ramesh, described the controversial claims as the boldest yet made by 'mavericks' working with stem cells in countries 'stretching from Mexico to China who cast themselves as crusaders against the medical establishment that has little to offer but comfort to terminally ill patients'. Researchers in Britain were cited as saying that stem cell therapies are decades away, and that, in both the West and the 'advanced economies of Asia', there are legal constraints that keep scientists from rushing ahead with treatments.

Two UK-based experts, Stephen Minger, the Director of King's College London stem cell biology laboratory, and Alison Murdoch, of Newcastle University Fertility Centre, were equally scathing in their criticism. Murdoch asserted that 'desperate patients might be tempted but false hope is no hope', while Minger stated that, 'It is highly implausible and frankly downright dangerous. If the Indian government wants to promote stem cell research then it needs to seriously look at regulation of these doctors and if necessary close them down.' The report proceeded to describe how 'western experts say that, potentially, she could be pushing an important therapy into the "realms of quackery", that her work is tantamount to "human experimentation"'. The message was clear; to be a global player, India needed to abide by the governance norms and ethical protocols established elsewhere, in this case one underscored by a Euro-American sensibility. That Shroff's 'maverick clinical trials' were well within her country's governance and legal protocols counted for nothing. The ethical and moral benchmark in the neo-Global order could not be predicated on an Indian legal sensibility that allowed novel and experimental treatment procedures on incurable and terminally ill patients (ICMR, 2000).

Surprisingly, while reporting positive experiences, and recovery in patients suffering from conditions as diverse as paralysis and motor neuron disease, the overriding tendency in the account is to either put any success down to 'belief', as opposed to rational, factual reasoning (Good, 1997)[2] or merely to a placebo effect. The experience of Poonam Singh, paralysed after contracting TB of the spine 11 years ago, is reported in a biased fashion. The patient experienced notable improvements in her condition, such as sitting up unassisted, feeling sensations in her legs, and even taking a few steps. This is described dismissively by the reporter in the following way; 'after a month of injections, she *believes* her condition is improving' (emphasis added). The irony is hardly lost on a careful reader that the only 'belief' being challenged by these dramatic experiences of healing claimed by patients is one in the rational and fact-based modality of scientific research. Also being challenged, through these dramatic admissions by patients, are beliefs in the professed investments in journalistic neutrality and scientific rationality (Dyke, 1995),[3] now confronted by seemingly 'implausible' and 'maverick' science. Throughout the article, a binary between good and bad science is posited by splicing the reportage of Dr Shroff's claims, and those of her patients, with expert voices from the United Kingdom. The non-compliant and unyielding local is repeatedly questioned by the global. In addition, a global benchmark is reasserted, as the reading publics are enrolled into an ideological and normatively binding 'value' of peer-reviewed good science that must apply as 'fact' to all scientific research across the globe. Under the neo-liberal global order, such standardization is the minimum requirement in order to produce the untainted research, development, production, and application of ESC entities. There is the creation of moral and ethical 'comfort zones' of scientific governance and scrutiny, which resonate with the regulatory and policy concerns in much of the European and North American landscape. In all these accounts, the agency of terminally ill patients, or severely disabled individuals, is belittled as 'desperation and hopelessness'. An encounter with unscrutinized science, in the view of vociferous critics, cannot be anything else.

Stem cell miracles and miracle stem cell cure

On 23 January 2006, the UK-based Sky News carried a prime-time feature on Dr Shroff's clinic and patients. The report was predictably entitled, 'The Stem Cell Miracles'. Sky News ran another feature on 13 April 2007, and imaginatively entitled that, 'Miracle Stem Cell Cure'. The only way, it seemed, that local and global media could make sense of purportedly 'maverick' science was by invoking the language of miracles that lay outside the realm of rational probability and scientific possibility.

The January 2006 report opened by stating that the clinic looks like an unlikely venue for a revolutionary treatment which could change the face of medicine worldwide. The establishment is described as the kind which someone might think twice about visiting, but 'inside is a doctor who may be – just may be – curing the incurable' (Sky News, 2006). Dr Shroff is described as 'someone who trusts few', who reluctantly agrees to be inter-viewed. The report states that her apprehensions had only strengthened after an interview with a British journalist (alluding to the *Guardian* report) who 'aimed to make her out to be a quack, an uneducated woman, who is experimenting on some of the most vulnerable people around – patients who have no hope'. However, in the interview, Dr Shroff responds to her critics by saying, '[M]y patients are my proof. None of them have suffered side effects. All of them are getting better. I will tell in time, my own time. It will be soon. They don't want to believe an Indian woman has managed to do what they have not' (Sky News, 2006).

In the news story, 25-year-old Poonam Singh was described by Dr Shroff as the 'star patient'. Singh, who had also featured in the *Guardian* news report, had received her first stem cell injection in August 2005, and states that, 'sensation began coming back to my legs, and just before Diwali [November] I took my first steps. I could not believe it, after so long. Now I really believe I will walk again. I am very happy' (ibid.). The Sky News Asia correspondent, Alex Crawford, proceeds to introduce a 'rather grumpy looking man', Mr Ajmera, who is suffering from Parkinson's disease. According to Crawford, he was diagnosed with the disease in 1995:

> [W]hen he came to Dr Shroff he was in a wheelchair, unable to walk, with severe rigidity in his arms and legs and so much shaking and loss of control of his limbs he could not do much for himself. I ask him how he is. 'I am fine', he answers holding out his hands to show me how steady they are. They don't move at all. Not even a whisper of a quiver. And then to our slight alarm, he jumps up. 'Look, I can walk' he twirls for the camera, 'I can sit', he sits for the camera, 'I can jump' he jumps for the camera, his arms spread-eagled.
>
> (ibid.)

The tone of the report hovers somewhere between disbelief and suppressed excitement at witnessing such 'miraculous' stories of healing and recovery. It attempts to strike a balance by citing dissenting voices and concerns, ranging from Stephen Minger to Raj Bahn, the Indian government's Biotechnology Secretary. The report concludes by quoting another paralysis patient saying, 'I don't care about what happens to me in the future. My life was not worth living before. I would much rather live this now, with some movement, for two years, than live like a vegetable for a hundred' (ibid.).

In the second Sky News report, Alex Crawford broke another sensational piece of news in April 2007. Sonya Smith, an Australian, who had been told that she would never walk again, was on her feet within weeks after receiving embryonic stem cell treatment from Dr Shroff. According to the news report, Smith had been told to prepare to spend the rest of her life in a wheelchair. After 18 months of relentlessly trying to get help, she flew to New Delhi in order to try out the treatment. After eight weeks of injections, Smith is reported as having regained full bladder and bowel control, her leg muscles are reportedly growing, and she can feel deep sensation in her thighs and feet. In the wake of this, Dr Shroff attracted renewed criticism, but insisted that she was doing 'nothing wrong' in following her own country's guidelines, and that her 'proof' was her patients (Sky News, 2007). The most dramatic assertion came from Crawford herself; 'We interviewed dozens of her patients and have spoken to many several times since [January 2006]. None has demonstrated any side-effects and all have shown signs of improvement in their respective conditions' (ibid.).

The British broadsheet, *The Daily Telegraph*, described how the T11 and T12 bones in Sonya Smith's spine were broken when she was crushed by her car as she tried to stop it from rolling down a hill after a handbrake failure (*Daily Telegraph*, 2007). In the newspaper article, Dr Shroff defended her therapeutic intervention, asserting that, 'Mrs. Smith has not had a "miracle", but she no longer has to suffer the indignity of catheterisation and wearing diapers' (ibid.). The now familiar criticism reasserted the moral, ethical, and irresponsible nature of such interventions, with Stephen Minger leading the chorus of disapproval, 'I think it's dubious to say the least. She might be Mother Teresa, but I am sceptical.' The suggestion from experts that this might be yet another instance of the placebo effect was quickly dismissed by the recovering Smith, 'it is definitely the stem cells ... I can feel my thighs where before I couldn't' (ibid.).

The Sky reporting could not be more different from that of the *Guardian* and *The Indian Express* pieces on Dr Shroff. While adhering to the journalistic canons of objective, disinterested and balanced representation, the two sets of reports show how journalists can become enrolled into stories that they report. In his reporting, Randeep Ramesh, of the *Guardian*, resurrects the binaries between good and bad science, First and Third World, expert and maverick. On the other hand, Alex Crawford remains sceptically on the fence, electing to report her daily experience of meeting the patients and interacting with Dr Shroff. In the former case a rhetorical device is used to discredit the subject of the story by invoking the well-established 'public fears' about mavericks, and in the latter case journalistic evidence is produced to fill the void left by the notable absence of scientific evidence and scrutiny.

Liminal envy? The desire for anti-structure

As already noted, the rise of India, alongside China, as the potential economic hub of the world, is being widely predicted and anticipated. These predictions/anticipations are often enunciated from global spaces as diverse as media, policy documents and political and scientific rhetoric. Scientific and political responses emerging from contemporary Britain are particularly interesting. Tony Blair, after his 2005 visit to India and China as the British Prime Minister, was widely quoted in the media as addressing his cabinet colleagues on those countries' achievements. His main contention was that the Labour government must secure Britain's future in a world of rapid change, driven by globalization, in which India and China led the way. Similar speeches in the context of the European Union are increasingly becoming the norm. These fears, however, are not confined to the political sphere.

On 7 March 2005, the BBC News website carried a story entitled; 'Is the UK Losing its Way with Stem Cells?' The report cited British scientist Professor Sir Chris Evans, a venture capitalist and bio-scientist, who argued that the UK could fall behind Asia if scientists were not able to make the step from research to clinical trials much more quickly. The president of the Royal Society, Professor Sir Richard Gardner, complained in the news report that researchers in Britain have been put in a straitjacket of regulation. He is quoted as saying:

> We are getting closer and closer all the time with research on heart disease and neurodegenerative diseases, but we are not helped by the amount of regulation. Other countries, where we have seen a lot of progress, do not face the same restrictions.

These two prominent scientific voices clearly articulated their displeasure with the high levels of regulation of science in contemporary Britain. Ironically, these comments were made at the same time as the Indian State was attempting to bring regulations governing stem cell research in line with the level of control seen in the Euro-American countries. Well-regulated science in Britain, on the other hand, was now craving the *liminal third* that enabled some Indian stem cell scientists to achieve 'breakthroughs'.[4]

On 24 March 2005, the UK Commons Science and Technology Committee Report on Human Reproduction Technologies and the Law was released. The report's chief recommendations were to have less regulation of IVF, to abolish the fertility watchdog, the Human Fertilisation and Embryology Authority (HFEA), to allow the creation of part human, part animal embryos for research, and to permit parents undergoing IVF to choose the sex of their baby. Earlier, in December 2004, British fertility expert Professor Lord Robert Winston had questioned HFEA's continuing value to fertility treatment. Speaking to the BBC, Lord Winston argued that he would like to see a more efficient body in the HFEA's place; 'something a great deal less bureaucratic, which doesn't inhibit research, which has a better consultation process with the public and which has a much more adequate inspection process ... it is time for Parliament to revise what's happening'. Little over three months later, the HFEA was being scrutinized in the Commons Science and Technology Committee Report. The proposed changes in the domain of human reproduction must also be seen in the context of embryonic stem cell research. The HFEA had played a central role in granting permission for therapeutic cloning of embryos to a team of scientists based in Newcastle University. The HFEA had come to be viewed as excessively rigid in the light of a few, high profile, and highly publicized, cases involving the creation of embryos to help seriously sick siblings. The timing of the Commons report had also been politically fortuitous for the scientific establishment, which felt encumbered by excessive regulation and desired less structural control over the way in which they 'do science'. Most importantly, these developments raise conceptual questions, such as: What is to count as global as opposed to local? A normative benchmark, or gold standard, ethical and legal protocol is simply unavailable in a globalized research system. If anything, the multinational context of scientific research has had unprecedented impact on what is to count as adequate regulation. More than ever before, the governance of science in the twenty-first century has a shifting focus, one that hybridizes with every new instance of the 'culture of research'. There was a time when the world was imagined as simple binaries, of the developed and the developing, the North and the South, the East and the West, underscored by a unidirectional flow of knowledge, resource and technologies. These binaries have collapsed, as nation–states, economies, science and technology compete and search for new markets, for a global level playing field. Little wonder the search for a level playing field is leading countries such as India and the United Kingdom to rethink the existing states of governance, ethical and legal controls in the neo-liberal mode of production.

Promissory visions

The news media in India are full of reports of the miracle cures of stem cell therapy, which is touted as the answer to many chronic ailments (Srinivasan, 2006). Some argue that it is unorganized and unregulated therapeutic stem cell *research* being passed off as therapy by its practitioners (Pandeya, 2007). The justification for continuing with this work is usually that the patients are desperately ill, and that even the limited success claimed by the practitioners is sufficient to provide hope for a miracle cure for the hopeless and the desperate (Gentleman, 2007). One early example, reporting the results of treating patients with a condition called aplastic anaemia, which causes damage to bone marrow and is fatal if the patient's blood is not regularly replaced through transfusion, was described by a UK-based newspaper as: 'No "voodoo magic." But maybe a miracle' (Lee, 2004). In what was described as an ongoing medical trial being conducted by Dr Ilham Abuljadayel in a Mumbai hospital, patients' cells were treated using a technology based on retro-differentiation that generates embryonic stem cells from adult ones. But, as noted in the *New Scientist*, little of this work appeared in peer-reviewed journals (Coghlan, 2003, 2004), even though, according to Lee, some patients exhibited evidence of regeneration.

The All India Institute of Medical Sciences (AIIMS) has been using stem cells derived from a patient's bone marrow and injected into an artery leading to diseased heart muscle to help cure heart disease. In 2005, AIIMS reported a trial in which all 35 patients survived and 64 per cent of their dead heart muscle was revived. The aim was to obviate the need for heart transplant surgery in the face of an acute shortage of donors. However, little of this work has been peer reviewed and indeed none of the research was authorized by a national ethics committee, only by the local committee in AIIMS (*Times of India*, 2005c). An editorial celebrating this 'success' in the *Times of India* looks forward to the possibility that India could become 'a hub for frontier medical treatment at competitive prices' (*Times of India*, 2005b). A leading clinical researcher goes further to suggest that India will soon become a world leader in stem cell research and regenerative medicine. Unsuccessful, lengthy and painful treatment will be replaced with off-the-shelf regenerative solutions (Cherian, 2006).

The few procedures in India that have been successful, as, for example, in the case of stem cell cures for blindness (Randerson, 2005), are held up as models by the so-called 'mavericks' who resist external peer review. Together, these stories construct a future when such cures are commonplace. This future is, however, being mobilized in the present to manage uncertainty and to organize resources for further work (Brown and Michael, 2003: 4). '[A] culture of innovation, driven by vision and hype, is not simply a waste or unreal but rather an extremely productive mechanism of value generation in a speculative marketplace' (Sunder Rajan, 2006: 110). The vision the stories seek to create also forms a *normative* space for enrolling allies from the wider culture, legitimated through the morality of providing help and succour to the poor and needy (Berkhout, 2006). As Borup and colleagues note, vision stories embody 'generative' expectations that:

> guide activities, provide structure and legitimation, attract interest, and foster investment. They give definition to roles, clarify duties, offer some shared shape of what to expect and how to prepare for opportunities and risks. Visions drive technical and scientific activity.
>
> (Borup *et al.*, 2006: 285)

Conclusion

It is widely recognized in India today that India's first IVF baby was born on 3 October 1978, 67 days after the birth of the world's first, Louise Brown. Dr Subhas Mukherjee announced the birth of Durga in Calcutta (Anand Kumar, 1997; ICMR, 2004). Dr Mukherjee is credited with many firsts in the world of reproductive and infertility medicine, including the use of gonadotropins for ovarian stimulation, the transvaginal route for harvesting oocytes, and the freezing and thawing of human embryos before transferring them to the uterus. Mukherjee published a short note on the birth in the *Indian Journal of Cryogenics* (ICMR, 2004), but his claim was largely discredited. Deprived of opportunities to pursue his research, humiliated and ridiculed by his peers, eventually Mukherjee took his own life. The maverick 'quack' was to remain unrecognized until the late 1990s, when his old diaries, papers and research data were resurrected. The circumstances leading to Mukjerjee's suicide, and the politics behind his subsequent reinstatement as the scientific 'father' of India's first and world's second test-tube baby, are analysed elsewhere (Bharadwaj, 2002).

Mukherjee's case is important as it draws our attention to established links between the purportedly opposed terrains of normatively compliant, rule-ordained good science, and that which is construed as anomalous, anti-structural bad science. The chapter has suggested that the dimension of *liminal third* might be a fruitful way to conceptualize the implosions of the seemingly antithetical categories into open-ended third spaces. This implosion produces a critique of what is official, normatively given, and powerfully structured. These critiques, departures, and/or transgressions, are important if we are to understand the workings of science on a scale where it is becoming increasingly difficult to speak in terms of local and global spatial distances. At the very core of a dualistic imagination is a spatial and temporal distance that late-, or post-, modernity has fatally eroded. In the example above, we find Mukherjee, positioned in 1978 in the *liminal third*, in a Third World and developing country, announcing a breakthrough that seemed ludicrous to his peers. However, Mukherjee lived and worked in a world preoccupied with the Cold War. In this world, it was possible to sustain the discourse of a privileged global North against which everything else was to be measured. In the contemporary world order that foundational belief is rapidly evaporating. However, as the case of Dr Geeta Shroff shows, the media debates on her work, which have cast her primarily as a 'maverick', have chosen to comment on her ESC research and application from a position of disbelief. She is pejoratively cast in the role of 'bogus miracle worker'. At one level, the disbelief and outrage evinced here is not too dissimilar from that encountered by Mukherjee, occasioned by an absence of peer review in a world in which research is driven by venture capital and robust state investments. In dramatic contrast to multinational, large-scale biotechnology research, both Mukherjee and Shroff worked in a 'third space' of a small, self-funded laboratory. And yet, the *liminal third* compels a constant reordering, whether by a so-called maverick scientist experimenting with embryonic stem cells in New Delhi, or by oppositional voices in the UK which fear the loss of competitive advantage in an over-governed research environment.

6 One ethic fits all?

Complexities in cross-cultural standardization, bioethics and regulatory protocols

Introduction

The United Kingdom has recently set up the world's first Stem Cell Bank. Only stem lines adhering to the Bank's stringent ethical guidelines and provenance protocols validating the human embryonic materials are admissible for donation. This development poses hitherto unexplored questions concerning the standardization of ethical and regulatory practices in a culturally diverse world. This chapter explores the way in which embryonic stem lines currently under development in India are responding to growing moves towards ethical, legal and regulatory standardization around the globe, and the way in which this process is seldom unidirectional. In so arguing, this chapter also shows how the bioethical framing of biotechnologies is shaped by local/global cultural processes and in turn shapes that cultural context. India has formulated draft Guidelines, based on the regulatory models of the UK and the USA, as part of the need to meet the growing regulatory concerns of countries that have invested heavily in stem cell research. This allows India to produce high quality laboratories staffed by well-qualified scientists and technicians, ensuring that the resultant embryonic cell lines meet the stringent provenance protocols in the UK and the USA. However, these Guidelines have yet to become legally binding, and working conditions for Indian scientists remain flexible enough to allow a small number of unregulated therapeutic interventions using stem cell technology.

This chapter begins with a discussion of the growing international cult of the 'maverick scientist', looking particularly at the rise and fall of Professor Hwang of South Korea. It suggests that the well-documented trade in human organs in India, and that it is used as a site for trialling new drug therapies, make it an ideal location for stem cell work. This is further contextualized by reference to the continuing biotechnology boom based on developing cheap generic drugs for Third World markets. This chapter continues by outlining public sector regulation of stem cell research in the United Kingdom and the United States, before discussing the process of adopting the draft Guidelines in India. The key area of informed consent is then analysed in an attempt, following Jasanoff (2005), to develop a deeper and more nuanced understanding of the ways in which normative discourses about biotechnology are embedded in wider culture and practice. The chapter concludes that India is able to foster leading edge research in therapeutic interventions by providing a liminal space within which to develop stem cell technology, whether by 'maverick' or by 'orthodox' scientists.

Maverick science?

In July 2006, a leading British broadsheet newspaper, reporting on the work of Panos Zavos, who runs a private fertility clinic in Cyprus, published a report entitled 'Maverick medic reveals details of baby cloning experiment' (*Guardian*, 2006). In an open letter to media editors, 13 scientists, including the President of the Royal Society, Lord May, said that maverick scientists were more interested in publicity than advancing the cause of science. Zavos was not the first high-profile scientist to be identified in public as a maverick. In 2004, the British sociologist Hilary Rose publicized the term 'cowboy cloners', a phrase she borrowed from Dame Suzi Leather of the Human Fertilisation and Embryology Authority. Rose was cynical about cloning's likely impact on improving human health, and wrote, 'Here we go again … There really needs to be phrase to describe this researcher's equivalent of the old charge against doctors of shroud waving' (Rose, 2004). She suggests that the lust for scientific glory may be more important to some scientists than the medical imperatives.

Her remarks came after the South Korean Professor Hwang Woo-suk announced his breakthrough in cloning patient-specific stem cells. However, a report in the *Guardian*, dated 24 December 2005, headlined, 'Cloning fraud hits search for stem cells', described how Hwang had been forced to resign from his senior post at the Seoul National University following the discovery that he had falsified at least nine of the 11 embryonic stem cell lines that he had announced in his paper in *Science*. Hwang's vision of creating a World Stem Cell Hub in South Korea, linking researchers in Europe and the United States with his laboratory, was shattered. After the announcement on 1 November 2005 that the Hub was open for patient applications, more than 3,500 patients jammed its phone lines and internet servers. The Hub's success would have ensured Hwang's, and South Korea's, leadership in stem cell research for the foreseeable future. The principal cause of Hwang's downfall was fraud and a failure of the peer-review process. However, an ethical scandal was also involved. Hwang admitted to using in his experiments eggs donated by a subordinate researcher in his lab. This admission followed reports that illegal eggs were being trafficked in his laboratory. This scandal brought to the fore the continuing debate about what constitutes 'informed consent', a principle that is the foundation of any ethically rigorous regulatory structure. South Korea had announced its desire to establish a leading-edge biotechnology industry in the 1990s, launching Biotech 2000 in 1994, and pledging to spend US$18 billion over a 14-year period. It aims to be one of the seven top biotech nations by 2010, despite a poor science base and a fragile biotech industry in which less than 10 per cent of pharmaceutical companies are involved in R&D activity. While other countries, such as the USA and the UK, addressed the need for clear and transparent regulation before permitting research to continue, South Korea allowed its scientists to forge ahead. Hwang was able to establish his laboratories in a regulatory environment where 'financial accountability, transparency, and regulation were sorely missing, and institutional walls against corruption, cronyism and other forms of misconduct weak' (Gottweis and Triendl, 2006: 142–3).

Stem cell technology research in India, as shown in earlier chapters, illustrates the similar difficulties faced by rapidly growing and ambitious economies with poorly developed regulatory agencies. India already has a well-documented, ethically contentious, local and global trade in human organs. Because of its population size and its inability to provide adequate medical care for its people, it continues to be vulnerable to being seen as a global locale for unethical or maverick science. It has been described as the 'guinea pig' of the world (Bound, 2007: 44). As one commentator put it: 'The days of the Raj are long gone, but multinational corporations are riding high on the trend toward globalization by taking advantage of India's educated work force and deep poverty to turn South Asia into the world's largest clinical-testing petri dish' (Carney, 2005).

There exists a steady supply of human gametes to the West, illustrated by newspaper headlines such as, 'Indian semen heads for Europe' and, 'Indian ovum finds world market'. The role of the religious nationalist-led Indian state in the recent past promoting the 'spectacular technology' view of bio-technology as part of its vision of a glorious Hindu past is perhaps the most significant, especially as its impact is similar to that of the South Korean state in the promotion of bio-technology. Stem cell technology has been targeted as a key vehicle to achieve the aim of establishing India as a significant player in biomedical research, using spare eggs obtained from the thousands of Indian infertility clinics offering in vitro fertilization (IVF) treatment. Informed consent is manufactured in the Indian context, as shown earlier, in part through ignorance and in part through collusion.

The biotechnology boom in India

The emergence of India as an emerging global economic superpower problematizes conventional assumptions about the division between First and Third World economies, particularly as India's strengths lie in the sophisticated IT and pharmaceutical sectors (Frew *et al.*, 2007). This suggests that the privileged uni-directional flow of information from North to South may now be an unsustainable project. Partly, this lies in the apparent 'paradox of Indian science', as noted in Chapter 1, analogous to the supposed paradox of the bullock carts and BMWs that jostle together on its crowded urban roads (Bound, 2007). The impact of IT on the Indian economy has been enormous, rising from 1.2 per cent of GDP in 1999/2000 to a predicted 4.8 per cent of GDP in 2005/6 (ibid.). IT has been challenged, as India has become a 'biotech hub' for the production of cheap generic drugs for sale in poorer countries. However, the ratification of WTO IPR Agreements is forcing Indian pharmaceutical companies and university departments to increasingly focus on innovative R&D and in making India a preferred location for global inward investment in the genomics sector. It aims to have sales of US$25 billion and a market capitalization of US$150 billion (ibid.). In 1986, the Department of Biotechnology (DBT) was established to create infrastructure and opportunity as well as to identify priorities for the sector. However, this has presented enormous cultural, educational, organizational and financial problems at all levels of Indian society (Dhawan *et al.*, 2005).

Nature (2005) devoted a 30-page 'Outlook' section to India's 'biotech boom' in 2005. As reported from an analysis on behalf of Goldman Sachs, India's economy is set to be the third in size, after the USA and China, by 2032, although there are enormous contrasts between rich and poor among its billion-plus population. However, it is suggested that if the opportunities now present, together with the new political impetus behind the rise of 'neo-India', are judiciously seized, a future to be proud of awaits. Not that this is an unproblematic future for the biosciences. While Indian scientists are more productive per dollar than their counterparts around the world, it is also suggested that 'India lacks a critical mass of competent scientists in basic and clinical research' (Jayaraman, 2005b: 495).

The rise of 'neo-India', discussed in Chapter 1, provided the necessary transformation from the postcolonial economic model required to facilitate the growth of directed market sectors in the 1990s, most notably in information technology and, more recently, biotechnology. The *New Scientist* in February 2005 described India as the 'next knowledge superpower', basing its analysis on a 'knowledge revolution' marked by the establishment in India of R&D labs from over 100 IT and science-based firms in the past five years. Along with China, India will account for 31 per cent of global biotech R&D in staff in 2006, up from 19 per cent in 2004, supplementing, if not yet supplanting, supply of staff in, for example, the United States (Crabtree, 2006). This is alongside the increase in recruitment of graduates, both locally and from the expatriate communities in Europe and the United States, attracted by the new opportunities on offer. The Indian Pharmaceutical Alliance calculates that some 10 per cent of new recruits at senior level are expatriate Indians or foreigners (Jayaraman, 2005a).

Underpinning these concerns is the need for a 'new way of thinking' that requires the building of an enterprise culture that brings science and industry together (Webb, 2005: 41). Some of this building will come from the expatriate scientists who have already been exposed to a more innovative western culture. Some will come from strengthening the large network of public sector research centres, and by introducing knowledge parks and genome valleys, with high quality infrastructure, simplified regulation and entrepreneurially oriented tax breaks (Jayaraman, 2005a: 481; Randerson, 2005a: 42). India is, according to the World Health Organization, already the fourth largest producer of pharmaceuticals, with two-thirds of its exports going to developing countries (Jayaraman, 2005a: 480). It is also the leading supplier of generic drugs, an industry developed before accepting the WTO IPR Guidelines and now maintained by producing medicines whose patents have expired (Wilson, 2005: 44).

There is not, at present, a central database listing all the clinical trials (conducted by researchers within and outside India) approved by the Drug Controller of India. Comparisons have been drawn between these developments and parallel developments in the organ trade (discussed earlier), and in the provision of quick and cheap medical procedures (for example, IVF treatment, hip replacements, and corneal grafts) to the global market. The Indian government has been supportive of medical tourism, though this has not been without criticism (Sengupta and Nundy, 2005), and one Indian state established a Council for Medical Tourism in 2004 in order to promote itself as a health destination for foreign patients (Mudur, 2004).

The hopes of many in India, as elsewhere, have been raised by the hype surrounding the public debate concerning hESC research and its possible therapeutic applications. This followed the 'mushrooming' of both private and public clinics in India using stem cells in a variety of treatments (Sharma, 2006: Table I, 46–7), without anyone knowing what clinical studies were being carried out or how they were being evaluated (Jayaraman, 2005). Two Indian government departments that fought for some time to establish their own guidelines finally agreed over one set to monitor clinical practice (Salter *et al.*, 2007). Calls have been made for the draft ICMR-CBT Guidelines, published in 2006, to become law so that a robust regulatory framework can be established. '[N]ow the time is ripe to consolidate the gains from the public debates and collective wisdom of symposia and devise a statute which will clear all uncertainties currently facing stem cell research in this country' (Basu, 2006: 1478). Regulatory frameworks for human stem cell research are already in place in most countries in Europe and North America, and the key elements of those in the USA and the UK are highlighted in the next section.

Public sector regulation of stem cell research in the UK and the USA

Public sector stem cell research in the United States of America, as noted in earlier chapters, is funded through the National Institutes of Health (NIH). Funding is given on the advice of the National Academy of Science (NAS), a non-governmental organization, through its Human Embryonic Stem Cell Research Advisory Committee. Ethical input to this committee is provided by the President's Council on Bioethics. In 2006, US$609 million was earmarked for stem cell research, which represents 2.14 per cent of funding on all health research. Following President Bush's 2001 ban, public funding can only be spent on the 'federally approved' hESC lines created prior to 9 August 2001, although there are no restrictions on privately funded research, except where individual states operate their own ban. There is no organized federal effort to engage private companies in research co-operation or the funding of investments. Federal funds cannot be used for the derivation or use of stem cell lines derived from newly destroyed embryos, the creation of any human embryos for research purposes, or the cloning of human embryos for any purpose (Ambrossi, 2006).

Opposition to this policy was at its highest during the 2004 Presidential election, during which the Democrats enlisted celebrities as well as scientists, clinicians and patients to renew the ethical debate. Following re-election, attempts to loosen restrictions have been thwarted by an unprecedented use of the Presidential veto, first in July 2006 on a Bill allowing only a modest extension of embryonic stem cell research, in that it would allow the use of frozen embryos already tagged for destruction (Schwartz, 2006; *Nature*, 2006). The following year, Bush concluded his Message to the Senate by rejecting the proposed 'Stem Cell Research Enhancement Act of 2007', stating that, since it would allow the destruction of embryos under 14 days old, he could not allow American taxpayers and the nation to be compelled to cross this moral line (Bush, 2007). However, at State level, Californian voters approved Proposition 71 in 2004. This earmarked US$3 billion in direct spending on stem cell research and regenerative medicine over a 10-year period and established the California Institute for Regenerative Medicine (Winickoff, 2006).

The situation in the United Kingdom is markedly different, in that it emphasizes comprehensive regulation in order to ensure that all research is done in an ethical manner, a process that provides legitimacy for all stake-holders. Overall responsibility for stem cell research governance rests with the Department of Health through its UK Stem Cell Initiative Panel (UKSCI), which advises the Human Fertilisation and Embryology Authority (HFEA), the regulatory agency that reports to the Secretary of State. Ethical input comes from the Department's Gene Therapy Advisory Committee (GTAC), which acts as the UK national research ethics committee. Government funding for stem cell research is co-ordinated by a cross-council committee led by the Medical Research Council (MRC) and the Biotechnology and Biological Sciences Research Council (BBSRC). Approximately £85 million was spent on stem cell research in 2006, which represents 3.15 per cent of the health research budget. Human stem cell research must be licensed by the HFEA, whether it is funded publicly or privately, and embryonic research is permitted on any embryo less than 14 days old. Embryos may be created for research purposes through somatic cell nuclear transfer, or researchers may use donated 'spare' embryos from IVF treatment. In the latter case, the decision to donate must first be discussed with nurses independently of any clinical decision, in order to ensure that appropriate consent is given. No financial incentives are allowed, other than the reimbursement of expenses or reasonable loss of earnings. Co-ordination with private funding is encour-aged by the UKSCI in the areas of clinical research and drug development

(Winickoff, 2006). Following the implementation of the European Union's Tissues and Cells Directive, the HFEA and the Human Tissue Authority (HTA, established in 2005) had been expected to form the Regulatory Authority for Tissues and Embryos (RATE) in 2008.

Globally, the UK has the most liberal but also the most regulated governance structure within which to develop stem cells for research and for treatment, and was one of the first countries to approve and formally regulate research into the use of embryonic stem cells. To facilitate this process, the Stem Cell Bank (SCB) was established in January 2003, located in the National Institute for Biological Standards and Control (NIBSC) in Potters Bar. It is the first publicly funded bank of its type in the world. Only laboratories licensed by the Human Fertilisation and Embryology Authority can develop stem cell lines in the UK for deposit in the SCB (Glasner, 2005).

The aim of the Bank, according to its First Report in September 2004, was to create an independent and competent facility to produce, test and reproduce existing and new stem cell lines derived from adult, foetal and embryonic human tissues. The Bank is a place to deposit adult and embryonic stem cell lines whose provenance has been carefully scrutinized and which have come from licensed laboratories. It has a commitment to ensuring that its cell lines are pure and free from contamination. In doing so, the Bank complies with the Department of Health Code of Practice for Tissue Banks (monitored bi-annually by the Medicines and Healthcare Regulatory Agency), its own Code of Practice (monitored by the Steering Committee of the Bank), and the European Union's Code of Good Manufacturing Practice (Stephens *et al.*, 2007).

A key question that the Bank's Steering Committee must address when accepting cell lines concerns the appropriateness and rigour of the procedures surrounding informed consent. Establishing provenance is essential to ensuring the legitimacy of the Bank, both in the UK and in the international stem cell community. The completed consent forms are part of the licensing procedures operated by the HFEA, whose representatives also sit on the Steering Committee, and are therefore not normally closely investigated when lines are deposited by UK laboratories. In cases where lines come from outside the UK, all relevant documentation is required, but, in practice, may not always be forthcoming. This requires a different kind of trust to be exercised by the Committee (Stephens *et al.*, 2008).

Cell lines are held under optimum storage conditions and are made available to other researchers around the world, on the understanding that any results from research using them will be publicly available. The Bank began to supply researchers in September 2006 (Sample, 2006), and by mid-2007, there were eight cell lines of research quality available for distribution, subject to the fulfilment of the stringent withdrawal requirements of the Bank. These requirements are potentially very bureaucratic, and the Steering Committee is considering accrediting certain research groups in order to facilitate the global movement of hESC material, focusing on the quality of the projects rather than that of the people. Requests for withdrawal from outside the UK, however, will still require thorough investigation of both projects and people (Stephens *et al.*, 2008). The Bank is thus concerned not only with the *husbandry* of cell lines, their production, stability, storage and monitoring, but also with their *identity*, their defining characteristics, to facilitate withdrawal by approved users. The staff of the Bank are aware of the contingent nature of both husbandry and identity in practice.

Adopting the Guidelines for stem cell research and therapy in India

In April 2005, a Stem Cells Workshop was held in Bangalore, organized by the National Centre for Biological Sciences, the Royal Society and the British High Commission. It was designed to foster communication, to allow scientists in India and the UK to exchange scientific knowledge and to explore opportunities for collaboration in the field of embryonic and adult stem cells. An overview of the official position in India was given by the Chair of the Stem Cells Taskforce of the Indian Department of Biotechnology (DBT), Professor Balasumbramanian. He described the road map for stem cell research, featuring an action programme focusing on clinical applications to be developed in city clusters such as Hyderabad, Vellore and Bangalore. It was agreed that the DBT should examine the UK HFEA regulations and guidelines with a view to establishing a similar mechanism in India. The DBT might also investigate the possibility of establishing a stem cells bank in consultation with the UK SCB. Subsequently, Stem Cell Research Forum of India (SCRFI), affiliated to the International Stem Cell Forum (ISCF), was set up to promote stem cell research, exchange ideas, disseminate scientific information, and provide public education, in an effort to find treatments and cures for a wide range of diseases. At its first conference in January 2007, it was announced that a public debate based on the draft Guidelines was to be held during that year in five Indian cities, in order to frame the rules and regulations covering ethics, good medical practice in manufacturing and clinical and laboratory research (*The Hindu*, 2007).

The draft Guidelines, based on similar ones in the UK and the USA, are clear that only supernumerary embryos can be used for research after informed consent is obtained from both spouses. These can only be collected in registered Assisted Reproductive (IVF) Clinics. Any international collaboration must obtain formal clearance from the state through a National Apex Committee (NAC), which will have responsibility for the scientific, technical, ethical, legal and social issues in the area of stem cell-based research and therapy. It will not, however, have regulatory authority over 'biologicals', which fall within the remit of ICMR, suggesting that some further clarification may still be necessary. The Guidelines also suggest establishing a Central Monitoring Committee, as a sub-committee of the NAC, to make site visits as required, and local oversight bodies to monitor particular institutions (ICMR, 2006).

In mid-2007, the Guidelines were yet to become law, and it remains unclear whether they will be followed assiduously when they do. Stories of hospitals undertaking procedures involving embryonic stem cell therapy abound. One commentator notes that these may well contribute to the ultimate success of India in establishing itself as a stem cell 'hub' since 'the Church in Western countries perceives using embryos to extract stem cells as murder. This, in conjunction with lax regulations on such research, has given India an obvious edge' (Pandeya, 2007). This view is confirmed by one of the respondents in this study, whose response to adopting Western ethical guidelines was to suggest that Indian science does not need Western ethical sermons as 'our embryos are not sacred'. In a survey of India's 179 institutional ethics committees undertaken in 2005, only 40 followed the existing ethical guidelines for research on human subjects (Mudur, 2005). Many private clinics that are offering stem cell therapy, according to the head of basic medical sciences in ICMR, Vasantha Muthuswami (quoted in Jayaraman, 2005), have never contacted the ICMR at all. Some of the numerous applications for funding have also been 'startlingly naïve', but Muthuswami believes that the checks and balances in the new Guidelines will ensure that hESC in India does not become a 'free-for-all'. This is in spite of comments such as those made by a surgeon at the prestigious All India Institute for Medical Sciences (AIIMS), who was reported in *The Times of India* (*Times of India*, 2005a) as saying that (then unpublished) clinical trials using autologous stem cells on 33 heart patients during by-pass surgery involved 'no ethical issues'. Muthuswami's response was enlightening:

> We are only a block away from AIIMS and we did not know that this was happening there. If the nation's premier medical institute did not ask our permission for such therapy, how can we blame private clinics for what they do?
>
> (Jayaraman, 2005b: 495)

A justification can be found in the words of a Western-trained Indian scientist who recognized the tensions that exist between the different cultural traditions, particularly in the contexts of commerce and social purpose that the new Guidelines reflect:

> All of us have some element of dual nationality. We are global citizens in terms of ethics and governance. We want to follow the best global standards. But when we step outside the lab, we become part of wider Indian society, which is more chaotic and occasionally corrupt. It's not straightforward.
>
> (Bound, 2007: 43)

In order to explore these contradictions further, a directorate member of ICMR was interviewed, Dr Raman. Dr Raman confirmed that India has adopted the UK Human Embryo Act, which allows specified research on human embryos up to 14 days of their creation. When asked how appropriate the Act was in the Indian context, Dr Raman accepted that the different communities within India would react differently, particularly with regard to understandings of informed consent. '[I]n the case of embryonic stem cell, the informed consent has to explain in detail, and you have to go by their religious belief, and those who do not accept it have to be given their freedom to decide.' Dr Raman went on to point out that similar problems face Western countries, citing Jehovah's Witnesses and their rejection of blood transfusions, and the similarities of these cases with, for example, the rejection of transplant surgery involving porcine heart valves by Jains in India:

> They'd prefer to die rather than accept a tissue of animal origin into [their] body, so it's the moral duty of the physician to explain to those individuals … they should be aware of their religious beliefs and religious compulsions … It's easier to get a porcine valve … But then these issues will come up. So in the stem cell transplantation, also it's very, very important … they have to be explained about the source.
>
> (Interview material)

However, the culture clash engendered by using Western notions of informed consent in the day-to-day work of Indian research regulation is revealed by the complicity of ICMR in accepting local practices that are at odd with global governance protocols. Dr Raman acknowledged that there were hospitals near to the ICMR headquarters in Delhi using stem cells from donated embryos of unknown provenance. Dr Raman went on to admit that the ICMR receive applications for permission to use stem cells from clinical researchers who expect to be allowed to carry out procedures the following day. Researchers ignore the requirements for information about the source of their stem cells, the way in which they are stored, the follow-up procedures that are involved, and so on. When asked whether these researchers are allowed to proceed, Dr Raman replied, 'They go ahead and do it. They are doing it.' Dr Raman cited the lack of any statutory framework for enforcing the Guidelines as the reason, but expects things to change after the establishment of the NAC:

> The only thing is you have to follow since we are giving all these guidelines, one has to follow the guidelines, and when such, any committee comes into existence, probably the committee will clear them. But at the moment, it is almost free for all.
>
> (Interview material)

However, Dr Raman also expects, in spite of evidence to the contrary, that the scientists involved in stem cell research will, out of self-interest, respect the official Guidelines during the consultation process:

It's, at the moment, it's not mandatory, it is voluntary acceptance of the guidelines to follow it. But it is expected that any scientists, in the proper sense of mind, will, would like to follow the guidelines so that they have put themselves in a safe position and not subject to criticisms.

(Interview material)

In Dr Raman's view, part of the difficulty may also stem from the restricted nature of the Guidelines' consultation process. Dr Raman felt that the wider public do not see their own input as being valuable when compared with that of journalists, activists and NGOs. The reasons for this include a historical lack of experience of public involvement in decision-making about science, and, simply, the lack of spare time to attend meetings such as the one held at the India International Centre in Delhi. We might conclude from this that many scientists and doctors in the clinical setting may deliberately ignore the Guidelines, safe in the knowledge that the existence of the Guidelines is unlikely to be something that their patients and non-specialist colleagues are aware of. The scientists and doctors may not even be fully aware of the Guidelines themselves. This appears to have been the case in China, where the journal *Nature* discovered an alarmingly low awareness of ethical regulations and informed consent procedures, even among researchers and medical professionals (Cyranoski, 2005: 138). This further highlights the problem of using Western models of informed consent in the Indian context. Bhutta noted that Western consent forms used in developing countries are usually translated and then back-translated to ensure that the original meaning is preserved. This literal emphasis serves to ensure regulatory legitimation and 'satisfy the legality of the process rather than the information and comprehension needs of the community' (Bhutta, 2004: 774).

Informing the consent process

One important aspect of ensuring that research on stem cell technologies achieves the legitimacy that allows it to appear as a neutral source of ethically validated stem cell lines, which have, by implication, no real history, is the means by which the processes of donation become standardized. In practice, the provision of information to potential donors of spare embryos for stem cell research varies between clinics in the UK, as it does elsewhere. Pilnick (2002) has argued that it is difficult to arrive at a meaningful notion of informed consent when seeking agreements to participate, particularly if the future uses of the donations are not clearly specified. This concern was addressed in detail by the recommendations of the 1999 Report of the US National Bioethics Advisory Committee. The Nuffield Council on Bioethics' discussion paper on Stem Cell Therapy suggested that the key issue for the United Kingdom is that of consent. The paper pointed out that the relevant Schedule requires that couples 'must be given a suitable opportunity to receive proper counselling about the implications of taking the proposed steps' and that they 'must be provided with such relevant information as is proper' (Nuffield, 2000: 9). In Australia, limited stem cell research, excluding reproductive cloning, has been legal since 2002 (Harvey, 2005). The process of revising existing guidelines on stem cell research to include therapeutic cloning has recently reached a conclusion following lengthy debate and public consultation (*International Herald Tribune*, 2006). The problem of obtaining meaningful consent was addressed by a major, government-sponsored review (The Lockhart Review, 2005), which noted the possibility of the coercion of young women donors through social disadvantage, family or workplace pressures. It suggested that strict guidelines were needed, and recommended that the Australian National Health and Medical Research Council should review these, especially with reference to any commercial potential or personal use of the products of such research.

The consent process needs to be contextualized by reflecting on the clinical encounter in which it occurs. Health professionals do not have a monopoly over the information required by patients in their decision-making. Research in the UK (Green *et al.*, 2006; Hundt *et al.*, 2006) suggests that decision-making is mediated by 'expert' interpretation, and that women hear certainty even when clinicians themselves are more circumspect in their opinions. This is exacerbated in the often hurried clinical encounters where procedures are explained, and results are analysed, simultaneously with the request being made for informed consent. Consent is not just about evaluating a particular technology, but also about assessments of trust and risk in professionals and institutions. Rose (2001) notes that the scandal in Alder Hey Children's Hospital and the Bristol Royal Infirmary in the UK, both arising from the collection and retention of body parts, may have contributed to undermining public trust in professionals and institutions. She also notes that medical records already in existence have often proved to be inaccurate or misleading, and no steps appear to have been taken to ameliorate this situation. Such concerns may be magnified when commercial organizations are involved. This is apparent in the controversies surrounding the introduction of genetic databanks in Iceland, Estonia and the United Kingdom (Martin, 2001; Sigurdsson, 2001). Societies such as South Korea and India may not be alone in finding that local interpretations of regulations concerning human subjects are messily contingent.

Legislative processes in India are governed by bureaucratic structures inherited from India's colonial past. However, when faced with novel techno-scientific developments, scientists and clinicians often find that these processes are cumbersome and slow even though they recognize the necessity for such processes. The British colonial system, as Dr Raman acknowledged in the interview, is good but difficult to maintain. This is highlighted by the process of obtaining informed consent for sourcing supernumerary embryos for research. The language of bioethics, as Hoeyer notes, may have universal pretensions but is 'in fact ethnocentric and incapable of accounting for moral complexity in a non-Western context' (Hoeyer, 2003: 241). It is wrong, says Dr Raman, to assume that all parties concerned are not fully aware of the apparent contradiction between appropriate regulatory procedures and the need to provide therapeutic interventions. Questions are asked and explanations are accepted, so 'people do understand very well' (interview material). However, the informed consent process needs to be contextualized in order to develop a more nuanced understanding of its application. Informed consent procedures constitute more than simply providing the basis for a rational judgement. They are also resources that can, for a variety of reasons, be left unused (ibid.).

A number of contextual issues in India arising from the interviews in this research provide the basis for just such a nuanced analysis. As documented, in earlier chapters, the extent to which patients in IVF clinics in India have a fiduciary relationship with their doctor, is partly because there is no national health service and many of the poorer patients cannot afford the treatments that they require. This situation significantly shapes the power relationship between doctor and patient and allows the possibility of spare embryos being donated as a form of payment for expensive fertility treatment. This situation is exacerbated by the traditional cultural acceptance of the moral authority of physicians.

In a study of patient encounters in rural India, Fochsen *et al.* (2006) note that these were dominated by the perceived knowledge and expertise of the medical practitioners, which was both undisputed and normative. Patients were seen as ignorant and incapable of understanding the information provided by the doctors, who regarded the patients' knowledge as misconceived or 'pre-formed' (ibid.: 1241). 'Patients ... see their doctors as next to gods', according to Dr Raman, echoing similar statements made by respondents in this study and elsewhere (cf. Bharadwaj, 2006a). Patients and doctors alike are not blind to these issues as being deeply culturally embedded and accepted. This is further complicated by the fact that India is a multi-lingual country with numerous dialects. Official consent forms, therefore, need to be translated into multiple languages and dialects, particularly in rural areas where literacy rates are very low. This makes the context in which translated documents are accessed and signed extremely problematic (Bhutta, 2004).

A distinction between 'spare' embryos and 'best' embryos occurs in IVF treatment, where judgements are normally made on clinical grounds to maximize conception. The use of spare embryos for stem cell research is not affected by this as the procedure for producing the embryos is the same. However, it can be linked to another important area that helps contextualize the debate about informed consent. This concerns gender discrimination, in which male children are preferred to female children in all aspects of life, from 'womb to tomb', but especially in terms of access to health care and education. Female infanticide is a common practice in parts of India, particularly in the South, with anything up to 500,000 girls lost in the womb or at birth every year, even though sex-selective abortions have been banned for more than a decade. This 'girl deficit' is more common among educated women, but does not vary according to religious belief (BBC, 2006). According to Dr Raman, '[I]t is said that if someone doesn't have a boy, then they cannot reach heaven, it's the boy that helps you to, leads you to heaven.' However, the motives underlying female infanticide may be more economic than cultural, rooted in the outlawed, but still practised, system that requires a woman's parents to pay the husband's family an expensive dowry at marriage.

This further highlights the importance of the extended family in relation to informed consent, particularly among Hindus, where in-laws often play a leading role in decision-making. This is exacerbated by the generally low status enjoyed by women and their lack of autonomy. As Fochsen and his colleagues suggest:

> Women were perceived as having an inferior status, which was, to a large extent, reproduced in the doctor–patient relationship. A woman would normally be accompanied to the doctor by either her in-laws or by her husband, if she was married, or by her parents if she was unmarried. During consultations, female patients appeared passive and submissive, and did not talk directly to the doctor.
>
> (Fochsen *et al.*, 2006: 1243)

Given the generally low status of women in India, medical intervention is one area where, according to Dr Raman, 'a man or woman always consults the family members which have a leading role to play'. As Dr Raman notes, consent is never really based upon individual autonomy, as in Europe or the United States, but is a collective family decision taken before the official consent forms are completed (Bhutta, 2004: 774). Occasionally, this throws up problems of confidentiality and privacy, as in the case of inherited diseases that could lead to discrimination, so such a process is not always either thorough or even practical.

Established informed consent procedures are one means by which a society can legitimate medical practices involving patients as ethical. Such procedures are intended to address issues such as the duty to protect life and health and to respect individual autonomy, through informed choice. Guidance, however, is often decontextualized from practice, and can result in informed consent becoming a ritualized formality. It appears that, in the case of India, guidance is not well established or closely followed. The reasons, discussed above, for this situation are not those that underpin the concerns of responsibility and trust that most Western ethicists focus upon, raising the possibility of 'ethnocentric prejudice' (Hoeyer, 2003: 241). However, without some form of legitimation, as Doyal (2004) notes, there is a danger of throwing the moral baby out with the critical bathwater.

Conclusion

The active drafting of bodies into the service of biotech futures in India can thus be understood in the context of: 'desperate', infertile couples, often from poorer social strata, in need of a quick and affordable resolution of their socially visible and debilitating condition; the use of informed consent by clinicians to mask the provenance of stem cell lines, and hide the wider socio-economic conditions from which they are derived; the vocabulary of altruism, of gift giving or donating, that reveals a hope by couples that today's act of charity will provide future benefit to others like themselves; and local regulatory regimes based on ethical practices developed within different cultural traditions and imported from Europe and North America. In this context, maverick scientists and clinicians can be seen as reflecting the deep cultural differences between India and societies with 'model' regulatory and governance structures, such as the United Kingdom or the USA.

There is a *moral economy* in which the ethical values of the global North – where embryos for research are in short supply and are embroiled in ethical and moral debates regarding their potentiality as sentient entities – support and encourage the donation of abundant, and ethically neutral, embryos by India and similar countries. Global advances in stem cell technologies, *shape*, and importantly, *become shaped by*, transnational, but locally based, collaborations comprising public and private research laboratories in India and abroad. India has formulated Guidelines on stem cell research, currently undergoing consultation, which are similar to regulations already operating in the UK and the USA. Their existence, while not legally binding, allows India to provide a supply of high quality laboratories, staffed by well-qualified scientists and technicians, while ensuring that working conditions for Indian scientists remain flexible enough to provide unrestricted access to the 'spare embryos' from its many IVF clinics. As a result, some Indian clinicians are using untested stem cell therapies, without appropriate ethical approval, in order to further scientific knowledge in this contentious field.

Guidance on informed consent is often *de-contextualized* from practice, which can result in it becoming a ritualized formality. It appears that, in the case of India, guidance, as understood in the West, is neither well established nor closely followed. Various dimensions of class, gender and education are potential sites for exploitation in the sourcing of embryonic and other bio-genetic materials, and lead to serious concerns regarding provenance. Stem cell technologies are still very new and raise complex bio-political and socio-technical questions, both within the UK and globally (Franklin, 2001; Waldby, 2002), and the scope for the future therapeutic application of stem cell research could be enormous. The contrast between clearly regulated research environments, such as those found, for example, in the USA, Australia and the UK, and those found in less well-regulated societies, such as India, could not be clearer. However, one unintended consequence of the strict regulation of stem cell research is the attraction that loosely regulated environments hold for scientists working at the frontiers of knowledge. For example, Singapore is rapidly developing its own group of international scientists (many from the USA) to work in this field (Schwartz, 2006: 1191). Likewise, until the downfall of Hwang, Korea had established the World Stem Cell Hub linking key US and UK scientists, in order to ensure its leadership in stem cell research and its medical applications (Gottweis and Triendl, 2006).

7   Conclusion

Introduction

In their Introduction to *Conceiving the New World Order* (1995), Faye Ginsburg and Rayna Rapp describe the uneven spread and take-up of bio-technologies, such as the new reproductive technologies, both within and between global locales. Their seminal intervention remains an anthropological landmark. Ginsburg and Rapp alerted us to the presence of entrenched *stratifications*, which offer differing opportunities, conditions, and circumstances, that enable and disable, and include and exclude women from experiencing reproduction and the technological benefits brought about by globalization. They reflected on the structuring of reproduction across social and cultural boundaries, particularly at the intersections of the local and the global (Ginsburg and Rapp, 1995). Furthermore, they contended that this is in no way a unidirectional process. On the contrary 'people everywhere actively use their local cultural logics and social relations to incorporate, revise, or resist the influence of seemingly distant political and economic forces' (ibid.: 3).

Local Cells, Global Science grapples with similar issues, albeit on a complex scale of finely woven interplay between that which is often simplistically conceived as the local in relation to a description of the global. The rise of embryonic stem cell research in India, and the global economic, political, ethical and legal circuits that frame these scientific transformations, provide a rich illustration of how agents use their 'cultural logics', resist when such situated logics are discredited by competing cultural articulations, and 'incorporate' as well as 'revise' science, technology and discourses about ethics and morality in pursuit of health, healing and profit. The *stratifications* encountered en route are often embedded in a hierarchy of global locales, institutionalized 'othering' tropes within science, legal and philosophical registers, and the location of agents in diverse social, cultural and economic spheres. That these *stratifications* are being 'scrambled' in the new century, as a new geopolitical order takes shape, makes the intersections of the local and the global not only contested sites but also third spaces, in which one can witness the rapid transformation and enactment of new, sometimes unprecedented, *stratifications*.

The appearance of India as an emerging global superpower makes problematic the conventional assumptions about the division between First and Third World economies, particularly as, recently, India's strengths have been in the sophisticated IT and pharmaceutical sectors. This suggests that the privileged, uni-directional flow of information from North to South may now present an untenable orthodoxy, and may be an unsustainable project requiring further study. Together with the political impetus behind the rise of 'neo-India' following the 1991 economic reforms, a number of issues for discussion by Indian social and natural scientists present themselves. The 'paradox' of India – that of the BMW and the bullock cart – conceals ideologically entrenched assumptions of an imagined Indian modernity, which is rooted in a bipolar, Cold War view of globalization. Commentators have been seduced by a vision of two Indias, and overlooked the creative possibilities that change brings about. The forces of change are being re-ordered in a novel *technoscape*, driven by a moral economy facilitating the rapid movement of information, knowledge, expertise and people. The introduction of genetically modified cotton and rice into Indian agriculture created world-wide scientific and media interest, and a substantial *indigenous backlash*. In contrast, the public debate surrounding India's burgeoning stem cell technology research programmes, which is often described as the 'next big thing to hit India after the country's software revolution', has been largely ignored.

The impact of IT on the Indian economy has been enormous, but it has recently been challenged as India has become a 'biotech hub' for the production of cheap generic drugs for sale in poorer countries. A recent focus piece by the Boston Consulting Group (BCG) described Indian domestic players in the pharmaceutical industry as having once 'antagonized multinational pharmaceutical companies', they were now attracting them in 'droves' (BCG, 2006). The recent ratification of WTO IPR agreements has forced Indian pharmaceutical companies and university departments to increasingly focus on innovative R&D, and on making India a preferred location for global inward investment in the genomics sector. However, this has presented enormous cultural, educational, organizational and financial problems at all levels of Indian society. India, because of its population size and its inability to provide adequate medical care for its people, continues to be vulnerable to being seen as a global locale for unethical or maverick science. It has been described as 'the guinea pig of the world', with parallel developments in the organ trade and in the provision of quick and cheap medical procedures (for example, IVF treatment, hip replacements, and corneal grafts) that serve a global market.

There is no robust ethical or regulatory structure in place to support the radical changes currently taking place in the red and green biotech sectors. In the case of genetically modified organisms, imports are permitted only for research purposes by GEAC (Genetic Engineering Approval Committee). Other uses fall under Guidelines, similar to those still under consideration for stem cell research, issued by the Department of Biotechnology's Review Committee on Genetic Manipulation (RCGM), which are open to various interpretations. While India boasts of becoming a global hub for stem cell research, there is no single government agency for licensing or monitoring stem cell research, and there is not, at present, a central database listing all the clinical trials (conducted by researchers within and outside India) approved by the Drug Controller of India. Ironically, while, to some extent, this renders India's position globally vulnerable, it also translates into a 'pull factor' that makes it an attractive destination for many. The central problem lies not so much in the perceived lack of regulation, as robust models are currently being enacted, and many are already in place in sectors such as drug trials and human subject research. Rather what remains problematic is that the proposed, and existing, robust regulations, allow the creation of merely 'risked' populations and subjects that can be seen as willingly sacrificing themselves for 'science', 'nation' and 'humanity' at large (Chapters 2 and 4). Furthermore, such subjectivity becomes the prime feature of the informed, autonomous, choosing and consenting neo-liberal citizenship. That such an enactment and the institution of regulation are both globally desired, and locally aspired to, in the hope of creating a standardized culture of research, development and application of science, is even more problematic.

As the book has shown, there is a *moral economy* in which the ethical values of the so-called 'global North' – where embryos for research are embroiled in ethical and moral debates – encourages the donation of abundant and ethically neutral embryos, by India and similar countries. Global advances in stem cell technologies, *shape*, and importantly, *become shaped by*, transnational, but locally based, collaborations comprising public and private research laboratories in India and abroad.

India has formulated Guidelines on stem cell research, which are currently undergoing consultation, similar to the regulations already operating in the UK and the USA. Their existence, while not legally binding, allows India to provide a supply of high quality laboratories, staffed by well-qualified scientists and technicians, while ensuring that working conditions for Indian scientists remain flexible enough to provide unrestricted access to the 'spare' embryos from its many IVF clinics. This liminal space allows some Indian clinicians to develop research in stem cell therapies, without bureaucratically defined ethical approval, in order to further scientific knowledge in this contentious field. Guidance on informed consent is often de-contextualized from practice, which can result in it becoming a ritualized formality. It appears that, in the case of India, guidance is neither well established nor closely followed. Various dimensions of class, gender and education are potential sites for exploitation in the sourcing of embryonic and other biogenetic materials, and must lead to serious concerns regarding provenance. In the end, the informed consent process is more or less as arbitrary in Euro-American and in Indian contexts. What makes these processes acceptable or unacceptable in either context, and determines the efficacy with which any national regulatory system can be enforced, is the perceived normative con-sensus on the degree of information that is available, on the 'imagination' of autonomous and individual citizenship, and, in the final analysis, the ability to rationally choose one strain of ethical thinking over another. Thus, as Chapter 3 shows, that the biosociality question in so-called 'non-Western settings' is situated between two contradictory ideological pulls, one pro-moting the notion of choice, and the other denouncing the very idea that a certain kind of choice be made. In other words, in the neo-liberal state of exception, it is never enough to just choose, but, rather, that choices must be made 'rationally'. Individuals embracing the stem cell therapies in India are repeatedly failing these rationality tests, and their biosociality remains unfulfilled, despite their bioavailability as the only ironic expression of their curtailed agency.

However, the potential for human exploitation, unfair trade practices, and morally contentious biogenetic research in the area of stem cell technology, is significant. This book has examined, for the first time, the transnational movement of tissues, stem cells and scientific expertise in the context of the nascent governance frameworks regulating research and development of biotechnology in India. Using these flows as a case study, the research has documented the impact that local and global ethical and governance frames have had on scientific practice and the everyday conduct of research in India. More importantly, the journey of 'spare' human embryos is traced from the point of conception in IVF clinics to public and private laboratories engaged in isolating stem cells, and developing potential therapeutic applications for local and global consumption. In so doing, the book has tried to demon-strate and understand the extent to which the global advances of the new biotechnologies shape, and become shaped by, transnational collaborations.

The research on which the book is based has also shed light on the way in which India is attempting to produce a stable regulatory order, with clear partitions and jurisdictions, within a global moral economy in the field of embryonic stem cell research. The process of doing so has generated co-constructed and messily contingent material hybrids (Brown *et al.*, 2006), and the book has reflected in some detail on the way in which their pro-duction is socially, culturally and materially shaped. Particular research regimes are enabled and facilitated by regulatory structures that also act to configure the innovatory outcomes. Decisions are made alongside the state in biosocial communities that construct narratives of neo-India through liminal spaces. India thus provides, in comparison with the USA and the UK, an interesting case study of civic epistemologies (Jasanoff, 2005), and of their role in democratizing scientific decision-making.

The relationship between science and democratic politics is undergoing a continuous process of change and co-construction. Nation–states are finding their sovereignty undermined by the global flows of labour and capital, multinational corporations and international social movements. However, announcements of the death of the nation–state may be premature. These global flows are heavily dependent on local economic conditions, and the protection provided by their home states, for their global successes. The states police these conditions to offer competitive advantages (low taxes, cheap and amenable labour, attractive interest rates, etc.) in the hope that touring capital will 'stop over' for longer than it takes to simply refuel. In wealthy democracies, moral responsibility is now something that needs to be paid for, rather than something seen as embedded in social and political institutions of the state – the citizen becomes the customer, with little regard for the costs elsewhere (Bauman, 1993). Morality becomes a commodity like any other.

In the main, the book has argued that India is being re-imagined and recast as neo-India in the twenty-first century through changes at all levels, including the local and the global. Change, however, is non-linear, and is resisted by entrenched structural constraints and imagined traditions which are perceived as incompatible with supposed modernities. Nor is change a new phenomenon for India, whose history is replete with encounters that have profoundly re-configured its culture. The book also argues that the local and the global are not merely sites to be journeyed to or from, but are fluid, and occasionally convergent, imagined spaces. Key stakeholders (clinicians, scientists, donors, families, and so on) inhabit these *dis-locations* while contingently carving a common ground for action. The dis-locations can be found at three levels: India as a rapidly emerging global economic player challenging its 'Third World' status; India as a site for innovative R&D rather than a cheap provider of technical labour; and India as a site for ethical research rather than maverick science. A key feature of this dislocating landscape is a neo-liberal *moral economy* in which more than simple commodities are produced. The guilt-free and sanitized circulation, consumption, and accumulation of ethically untainted commodities are also possible. However, their production, through agency, may be hindered as no ethical or moral consensus exists in the field of stem cell technology, even though a moral economy, defined outside India, prevails. Thus, the concept of *bio-crossings* describes those areas in biotechnology where the neo-global requirement for ethical consensus and the force of local, situated, practices meet. It is suggested that those involved in bio-crossings travel through unpredictable and open-ended *liminal third* spaces that subvert conventional polarities of local and global, maverick and scientist. In examining these unstructured third spaces surrounding embryonic stem cell research in India, in the biographies of Indian scientists, clinicians, patients and regulators, a more dynamic, yet nuanced, understanding of the civic epistemologies surfaces.

Civic epistemologies and the case of India

On the basis of recent studies comparing different innovation cultures, it is increasingly becoming clear that regulatory structures facilitate, and are constituted by, the innovatory process. In the biosciences in particular, attempts to introduce clear and stable boundaries, jurisdictions and regulatory orders cannot be seen as simply a linear process. Jasanoff, in her book *Designs on Nature* (Jasanoff, 2005), develops the concept of 'civic epistemologies' to provide a more sophisticated analysis, required to do justice to the richness and ambiguity revealed by comparing the development of biotechnological innovation in three western democracies (the USA, the UK and Germany). She focuses on the way in which debates are framed in order (or not) to present their subject-matter as political issues, the way in which what counts as science and what as politics or morality is established, and the question of who the actors are, and the discourses and kinds of reasoning they deploy. Her unit of analysis is the nation–state, organized around a dynamic concept of political culture rather than simply political actors and their interests.

Three examples form the empirical basis of her analysis: GM foods, embryo research (specifically technologically assisted reproduction), and patenting:

> Why, for instance, have agricultural biotechnology and [genetically modified] food not become openly controversial in the United States or Germany but do turn into matters of intense concern in Britain? How, to the contrary, did Britain succeed in carving out a relatively uncontested space for embryo research, while American politics on this issue remains deeply divided, and Germany refused to allow the most difficult choices to rise to political salience in the first place? Why is patenting life forms seen as an ethical issue in Europe, but not the United States? And what accounts for the fact that bioethics, simultaneously and energetically embraced as a policy discourse in the EU and in three sovereign nations, nevertheless is understood in vastly different ways in each of its contexts of development?
>
> (Jasanoff, 2005: 9)

Jasanoff discusses how, in Britain, an embryo in its first 14 days of gestation is described as a pre-embryo on the grounds that until then the cells would not yet have developed specific functions. It was this understanding, given scientific, political and religious authority by the House of Lords, which facilitated the passing of an Act allowing embryo research in the UK. In Germany, this research was banned after a similar debate, and special protection was accorded supernumerary embryos left alive after IVF to ensure that no human life was endangered. In the USA, it is usually the courts rather than federal bodies that provide the regulatory framework, allowing the intrusion of social and economic factors as well as scientific ones. The exception was President Bush's intervention, at federal level, into the stem cell debate described earlier in the book. However, the importance of morality and economics over science as the basis for regulation is startlingly clear. The justification for banning federal funding rests on ensuring that the American taxpayer is not involved in the 'death of human embryos'.

Current debates about biotechnology require, according to Jasanoff, not only an understanding of present scientific and political cultures, but also of preceding debates, the roles of different actors (including scientists and politicians) in these debates, and the ways the debates are framed through discourse. Science is effectively being constituted by these debates. Scientific claims become claims like any others, including those of the public and other stakeholders, and depend on their ability to facilitate the framing of a debate in order to make an impact on its outcome. Actively taking on board the involvement of citizens in the production, use and interpretation of knowledge for public purposes, constitutes a civic epistemology – a public way of knowing. It is a set of institutionalized practices by which members of a given society test and deploy knowledge claims used as the basis for making collective action.

To understand how we, as societies, know what we know, Jasanoff examines styles of public knowledge making, bases of trust or accountability, the way in which objectivity is demonstrated, and the foundations of expertise. National differences can be discovered in the foundations of establishing scientific expertise. In the United States, says Jasanoff, expertise is formal, and authority is based on published work in peer-reviewed journals. Representing particular interests, for example, environmental or commercial, can result in a loss of credibility. In the United Kingdom, being a technical expert is to establish credibility, as part of the expert function is to define the public good so that appropriate technical knowledge can be made available. In Germany, expertise is less obviously personal, and there is more emphasis on institutional backing from a variety of interest groups. Jasanoff concludes that different political cultures respond to different sources of authority, frame and understand seemingly straightforward technical issues in very different ways, and embed those frames in institutions that are not wholly responsive to changed understandings.

Jasanoff (2006) applies some of these concepts to India in her discussion of the power of social movements. She uses the example of the use of long-standing protests against the Government of India's plan to build numerous, often very large, dams on the Narmada River and its tributaries, in order to provide much-needed water and electricity for the drought-prone areas of Gujarat, Maharashtra and Madyha Pradesh. The protest in early 2006, led by the activist Medha Patkar, focused on the plan to raise the proposed height of the largest dam by a further 10 metres. Such movements and protests are a more common way of finding a voice for the public in India than, for example, in the USA or the UK. While in the UK, forms of questioning in policy are often embodied in expert judicial enquiries, and in the USA, with its adversarial approach, are resolved through expert testimony in the courts, in India they are also resolved through social protest rooted in authentic experience. The continuing history of the introduction of GM foods and crops into India provides ample testimony to support this suggestion. In 1998, for example, farmers in Andhra Pradesh and Karnataka torched transgenic cotton developed by the US multinational Monsanto, leading to the formation of the KSSR (Karnataka Rajhya Raitha Sangh), a farmers' movement in Karnataka which had a claimed membership of 10 million people by 2000 (D'Monte, 2000).

While the account of civic epistemologies advanced by Jasanoff recognizes the importance of nations and institutional settings, it does not do the same justice to their interactions with those local/global structures that establish the frames within which they operate. Nor does it allow for societies where public life is only a proving ground for competing knowledge claims in certain circumstances. The different responses to the introduction of GM food and crops, on the one hand, and the development of stem cell research in India, on the other, are examples where frames restrict access and define legitimate involvement. Jasanoff's account does not fully incorporate the fluid and dynamic co-construction of the third spaces that inhabit the bio-crossings in the moral economy. It is here where change is possible, but it is also where institutional framing competes to subvert social action. In the case of the farmers' rejection of GM crops, there are doubts raised regarding the question of whether the KRRS is indeed a mass movement, given that most of the population of Karnataka is made up of illiterate peasants, farming on scraps of land, whose knowledge and understanding of GM crops may be limited. The movement has often outflanked traditional political parties, underlining the shortcomings of existing democratic structures when dealing with biotechnological innovation. It is also the case that, when three-quarters of the population live in the countryside, it is relatively easy to whip up passionate support against technologies associated with India's former oppressive rulers or some foreign 'other'. However, as already suggested, stem cell research in India has not led to similar protests, perhaps disguising other factors that contribute to an alternative interpretation of civic epistemology. What kind of conceptual elaboration, therefore, adequately unpacks the experience of local and the global in the production, consumption and circulation of embryonic stem cell entities, in India, and in relation to other global locales? It is to some of these vexed questions that we must finally return.

Local and the global: further questions and reflections

The book opened with a reflection on Ashis Nandy's contention that 'in the name of science and development one can today demand enormous sacrifies from, and inflict immense sufferings on, the ordinary citizen. That these are often willingly borne by the citizen is itself a part of the syndrome' (Nandy, 1996: 1). When re-read in the context of the competition-driven, market model of neo-liberalism in the new century, an altered reading of Nandy's contention was proposed; In the *neo-liberal mode of production*, a citizen's 'willingness' to bear suffering is irretrievably distorted by socio-economic inequalities, just as the state's ability to demand sacrifice is enabled by the availability of gendered, stigmatized and impoverished citizens. This is as true of Euro-American countries as it is true of emerging neo-liberal states such as India. The moral economic thinking places an added premium on the ethical sourcing of materials, goods and services from around the globe in ways that raise a series of structural and economic issues concerning child labour, carbon footprint mapping, or fair trade coffee and food consumption. Morality becomes a constant and real economic 'cost' in pursuit of an 'ethical life'. For example, offsetting carbon emissions by signing up for, say, a tree plantation campaign in the 'Third World', or rescuing a polluting industry by using 'greener' technology, might facilitate guilt-free holidays or business flights, but does little to contribute possible solutions to the persistent and fundamental global inequalities that produce such pollution and 'polluted lives' in the first place. The only true 'moral' response in such a situation lies in planting trees or introducing greener technologies while being ignorant of their situated contexts. This, however, is an untenable option under the global neo-liberal moral economy. This neo-liberal moral and ethical synthesis arose primarily as a response to global political mobilizations against injustices and exploitation. It is threatened by the rapidly dis-locating global sphere, where the twentieth-century orthodoxy that viewed Global as Western and Local as non-Western has begun to rupture. However, the question of the West remains, both conceptually and theoretically, a difficult geopolitical and philosophical issue. To conclude, therefore, let us return to Ashis Nandy to (re)conceptualize the foregoing with the view to posing questions for further critical reflection.

In the Introduction to one of Ashis Nandy's anthologies, *The Return from Exile* (Sardar, 2004), which incorporates three of Nandy's notable books – *Alternative Science, Illegitimacy of Nationalism* and *The Savage Freud* – Ziauddin Sardar describes Nandy as someone who is not in favour of 're-engineering the old-fashioned, traditional, but somewhat world-weary *hindustani*', since the 'less than masculine' and 'scientific' Indian (as fashioned in the heydays of the Raj) has 'survived centuries of colonization and decades of modernity and instrumental development'. Even in the face of 'all-embracing embrace of post-modernism and globalisation' the *hindustani* seems to demonstrate a 'stubborn resilience' (Sardar, 2004: 1–2). Sardar proceeds to introduce the reader to that which Nandy is seeking to reject in his book *Alternative Science*: the dominant mode of western science. Nandy's proposed alternative is to a Western cosmogony that 'believes in the absolute superiority of the human over the non-human and the sub-human, the masculine over the feminine, the adult over the child, the historical over the ahistorical, and the modern or progressive over the traditional or the savage' (Nandy, 1995, in Sardar, 2004). This, however, must not be read as one alternative to the category of the West, or in any narrow sense, anti-West, for, as Sardar notes:

> '[for] Nandy, the West is more than a geographical or temporal entity; it is a psychological category' (ibid.). In this schema, the West is

everywhere: within and without the West, in thought processes and liberative actions, in colonial and neo-colonial structures, and in the minds of the oppressors and the oppressed – the West is part of the oppressive structure; it is also in league with the victims. Thus to be anti-West is itself tantamount to being pro-West.

(ibid.)

Sardar argues that:

> Nandy's alternative then is beyond the West/anti-West dichotomy, even beyond the indigenous constructions of modern and traditional options, in a different space. It lies in a new construction: a victim's construction of the West, a West which would make sense to the non-West in terms of the non-West's experience of suffering ... Nandy's alternative is the alternative of the victims; and whenever the oppressors make an appearance in this alternative, they are revealed to be disguised victims 'at an advance stage of psychosocial decay'. The construction of their own West allows the victims to live with the alternative West 'while resisting the loving embrace of the West's dominant self' ... There is thus no need to look elsewhere for an alternative social knowledge that is ethically sensitive and culturally rooted for it is already partly available outside the framework of modern science and social sciences – in those who have been the 'subjects', consumers or experimentees of these sciences.

(Sardar, 2004: 4–6)

Nandy's thesis is bold and provocative. The provocation does not lie in its embrace of the subaltern, for Donna Haraway, among many others, has shown that 'from below the brilliant space platforms of the powerful the vision is better' (Haraway, 1991; cf. Campbell, 2004). Subjugated standpoints are preferred because they seem to promise more adequate, sustained, objective, and transforming accounts of the world (Haraway, 1991: 191). Nandy's provocation lies in his attempt to dismantle the binary of the West and non-West by simply stating that India can never be the non-West. The 'Indian' cannot be resurrected as a counter-point to the West, an anti-thesis, nor can the Indian be romantically located in the 'suffocating traditions', or for that matter those traditions reanimated 'under the impulse of modernity' (Sardar, 2004: 25). Most importantly, his provocation leads to a confrontation with the, as yet, overlooked psychological underbelly of the dis-locating globe where 'West is everywhere'. And because it is everywhere, as material culture and as a philosophical register, as ideological dogma and as a normative belief, as brute oppressor and as a compassionate ally, and as perpetrator and as victim, it is in equal measure resisted and embraced.

Perhaps it is fair to ask what might the victims in the new century look like, and how might their imagination of the West define their existential location in the world? For instance, the rumours, in the Brazilian market town of Bom Jesus de Mata, of abduction and mutilation of children by Americans, to be used as fodder for spare parts, turned out to be just rumours (Scheper-Hughes, 1992). But, as Nandy shows, in the imagination of a victim, 'the construction of their own West allows the victims to live with the alternative West' as a measure of maintaining their own sanity and humanness. How might victims' constructs of the West then shape their survival strategy and hope for the future, and fashion the theodicies that help negotiate the encroaching global, the expanding local, and the contracting globe itself? More importantly, as Veena Das (2004) states, there are

> Neither pure victims nor noble resisters but a series of partnerships through which state and community mutually engage in self-creation and maintenance. This does not mean that we cannot engage with questions of justice and rights or that communities formed through suffering are delegitimized. The place from which these engagements occur, though, is not that of moral space of innocent victimhood but of the rough-and-tumble of everyday life
>
> (Das, 2004: 251)

Is it then possible to imagine the creation of a 'West' from a victim's *perspective*, contained within the mundane intricacies of the everyday, as opposed to *victimhood* predicated on struggle, surrender, and survival?

What lies at the other end of the spectrum? The reviled 'maverick' scientists in India, with their 'desperate irrational' consumers of embryonic stem cells, and the 'weak soft' Indian state (as once described by Gunnar Myrdal), unable to legislate the biotechnology sector because different levels of ethically sensitive, and culturally rooted, responses lie outside the framework of modern science and social sciences? Could the embryo donors, patients, or 'maverick' scientists in some direct or indirect way be the *failed* 'victims', 'subjects', 'consumers' and 'experimentees' of a 'Western science', since the science they embrace is criticized and 'othered' as failing to be *scrupulously* 'Western'? Is the maverick, then, a global category in a local world, ascribed to those refusing the 'authentic experience' of the 'West' and 'Western Science'? If so, can science only ever be just 'Western', the other of that which it seeks to colonize? Finally, given the burgeoning bio-crossings, is a different 'West' within the 'West' on the ascent, a West willing to embrace the 'inauthentic', 'maverick', and the 'dangerous other' in search of what Nandy so passionately describes as the *alternative*?

Notes

2 Dis-locations: local cultures of cells, global transactions in science

1 Biotech products include human and animal healthcare products, agri-biotech (including seeds) and industrial products such as enzymes, bio-instrumentation and bio-process equipment.
2 The English translations of ancient legal codes used in this book are not referenced as, first, these texts are not dated, and, second, they are subsumed under Müller's and other oriental scholars' 1894 translations. See Müller (1894).
3 *Upanishad* means receiving knowledge while sitting near the teacher, conversations between enlightened souls and their students on the subject matter. Upanishad philosophy is a critical reflection on the Vedic literature (there are 108 Upanishads) and seeks to understand the personal character of the Absolute Truth (http://www.vahini.org/glossary/u.html).

5 'Miraculous stem cells': the liminal third space and media rhetoric

1 The paper was based on 15 years of research, using data collected from 60 patients that demonstrated that the abdominal wall could be regenerated using a synthetic mesh sandwiched between two layers of membrane containing stem cells.
2 Byron J. Good asks how, historically, within Anthropology the central issue in the rationality debate remained focused on the question: How do we make sense of cultural views of the world that are not in accord with contemporary natural sciences? Good shows how, within anthropology and philosophy, the juxtaposition of 'belief' and 'knowledge', and use of belief to denote counter-factual assertions, has a long history (Good, 1997: 10–21). While Anthropology, especially post-war Anthropology, has actively problematized these issues, other disciplines, ranging from journalism to science, have remained mired in the dualistic and oppositional world of belief and fact, of rational and irrational.
3 According to Dyke:

> I regard journalistic accounts of science as autonomous acts of persuasion strategically phrased to disseminate a particular view ... Journalism and medicine are both self-regulating professions, defending editorial and clinical freedom, but largely depending on market fluctuations. Most significantly, the ideal of objectivity is inscribed in both discourses, and has materialized in professional routines, normative practices and textual conventions.
>
> (1995: 44)

4 A little less than a month from the date of the BBC news story, the British High Commission's Science and Innovation team, together with the Royal Society and the Indian government, organized an Indo-UK Stem Cell Workshop. The workshop ran from 4–11 April 2005 and was based at the National Centre for Biological Sciences in Bangalore. The Indian scientists involved in this venture worked mainly in publicly funded research facilities, and the workshop list of faculty did not include any of the private sector 'maverick' scientists that the news media routinely sensationalized (see Chapter 6).

Bibliography

Agamben, G. (1998) *Homo Sacer: Sovereign Power and Bare Life*. Stanford, CA: Stanford University Press.

Ambrossi, G.G. (2006) *Comparing Stem Cell Research in the US and UK*, Washington, DC: Center for American Progress. Available at: www.americanprogress.org/issues/10/pdf/uk_stem_cell_factsheet.pdf (accessed 12 August 2007).

Anand Kumar, T. C. (1997) 'Architect of India's first test-tube baby: Dr Subhas Mukerji (16 January 1931 to 19 July 1981)', *Current Science*, 72: 526–31.

Ansari, K.M. (2002) 'Indian semen heads for Europe', *Hindustan Times*, 3 May.

Appadurai, A. (1996) *Modernity at Large: Cultural Dimensions of Globalization*. Minneapolis: University of Minnesota Press.

Atkinson, P. (1990) *The Ethnographic Imagination: Textual Constructions of Reality*. London: Routledge.

Azariah, J. (1997) 'Status of human life in/and foetus in Hindu, Christian and Islamic scriptures', in J. Azariah, H. Azariah and D.R.J. Macer (eds) *Bioethics in India: Proceedings of the International Bioethics Workshop in Madras*. Available at: http://www.biol.tsukuba.ac.jp/~macer/index.html

Bakhtin, M. (1984) *Rabelais and His World*. Bloomington, IN: Indiana University Press.

Banerji, D. (1974) 'Social and cultural foundation of health service system', *Economic and Political Weekly*, 9: 32–4.

Bartholomew, S. (1997) 'National systems of biotechnology innovation: complex interdependence in the global system', *Journal of International Business Studies*, 28(2): 241–66.

Basu, S. (2006) 'Regulating stem cell research in India: wedding the public to the policy', *Current Science*, 90(11): 1476–9.

Battaglia, D. (ed.) (1995) *Rhetorics of Self-Making*, Berkeley, CA: University of California Press.

Bauman, Z. (1993) *Postmodern Ethics*. Oxford: Blackwell.

Bawden, T. (2006) 'Warning to British firms as India's UK investment rises', *The Times*, 17 January.

BBC (2006) 'India "loses 10m female births"', BBC News, 9 January. Available at: www.news.bbc.co.uk/go/pr/fr/-/1/hi/world/south_asia/4592890.stm (accessed 15 August 2007).

BCG (2006) 'Harnessing the power of India: rising to the productivity challenge in biopharma R&D', *BCG Focus*, May. Available at: www.bcg.com.

Berkhout, F. (2006) 'Normative expectations in systems innovation', *Technology Analysis and Strategic Management*, 18(3/4): 299–311.

Bharadwaj, A. (2000) 'How some Indian baby makers are made: media narratives and assisted conception in India', *Anthropology and Medicine*, 7: 63–78.

—— (2001) 'Conceptions: an exploration of infertility and assisted conception in India', PhD thesis, University of Bristol, United Kingdom.

—— (2002) 'Conception politics: medical egos, media spotlights, and the contest over test-tube firsts in India', in M. C. Inhorn and F. van Balen (eds) *Infertility Around the Globe: New Thinking on Childlessness, Gender, and Reproductive Technologies*. Berkeley, CA: University of California Press.

—— (2003) 'Why adoption is not an option in India: the visibility of infertility, the secrecy of donor insemination, and other cultural complexities', *Social Science and Medicine*, 56: 1867–80.

—— (2005a) 'Diffracting reproduction: infertility encounters stratified reproduction', paper presented at the American Anthropological Association 104th Annual Meeting, Washington, DC, 30 November–4 December.

—— (2005b) 'Cultures of embryonic stem cell research in India', in W. Bender, C. Hauskeller and A. Manzei (eds) *Crossing Borders: Cultural, Religious and Political Differences Concerning Stem Cell Research*. Munster: Agenda Verlag.

—— (2006a) 'Clinical theodicies: the enchanted world of uncertain science and clinical conception in India', *Culture, Medicine and Psychiatry*, 30(4).

—— (2006b) 'Contentious liminalities: embryonic stem cell research in India and the UK', paper presented at BIOS Seminar Series, London School of Economics and Political Studies, London, February.

—— (2006c) 'Reproductive viability and the state: the rise of embryonic stem cells in India', paper presented at Reproduction, Globalization and the State. Rockefeller Foundation Bellagio Study and Conference Centre, Italy, June.

—— (2008) 'Biosociality and bio-crossings: encounters with assisted conception and embryonic stem cells in India', in S. Gibbon and C. Novas (eds) *Biosocialities, Genetics and the Social Sciences: Making Biologies and Identities*. London: Routledge, pp. 98–116.

Bharadwaj, A. Atkinson, P. and Clarke, A. (2006a) 'Classification and the experience of genetic haemochromatosis', in P. Atkinson, P. Glasner and H. Greenslade (eds) *New Genetics, New Identities*. London: Routledge.

Bhattacharji, S. (1990) 'Motherhood in ancient India', *Economic and Political Weekly*, 20–27 October, pp. WS50–WS57.

Bhutta, Z. A. (2004) 'Beyond informed consent', *Bulletin of the World Health Organisation*, 82(10): 771–8.

Blakemore, C., Patel, N. and Gillespie, S. (2006) 'No cutting corners in stem cell research', *The Times*, Letters to the Editor, 29 August, p. 14.

BMJ (2001) 'India to tighten rules on human embryonic stem cell research', 323: 530.

Boerma, T. (1987) 'The liability of the concept of a primary health care team in developing countries', *Social Science and Medicine*, 25(6): 747–52.

Borup, M., Brown, N., Koonrad, K. and Van Lente, H. (2006) 'The sociology of expectations in science and technology', *Technology Analysis and Strategic Management*, 18(3/4): 285–98.

Bound, K. (2007) *India: The Uneven Innovator*. London: Demos.

Brown, N., Faulkner, A., Kent, J. and Michael, M. (2006) 'Regulating Hybridity: Policing Pollution in Tissue Engineering and Transpecies Transplantation', in A. Webster (ed.) *New Technologies in Healthcare*, Basingstoke: Palgrave Macmillan, 194–210.

Brown, N. and Michael, M. (2003) 'A sociology of expectations: retrospecting prospects and prospecting retrospects', *Technology Analysis and Strategic Management*, 15: 13–18.

Brown, N. and Webster, A. (2004) *New Medical Technologies and Society: Reordering Life*. Cambridge: Polity Press.

Buctuanon, E. M. (2001) 'Globalization of biotechnology: the agglomeration of dispersed knowledge and information and its implications for the political economy of technology in developing countries', *New Genetics and Society*, 20(1): 25–41.

Bush, G. W. (2007) 'Message to the Senate of the United States', Washington, DC: The White House Office of the Press Secretary. Available at: www.sba-list.org/files/2007-Veto.pdf (accessed 14 August 2007).

Campbell, K. (2004) 'The promise of feminist reflexivities: developing Donna Haraway's project for feminist science studies', *Hypatia*, 19(1): 162–82.

Carney, S. (2005) 'Testing drugs on India's poor', *Wired*, 19 December. Available at: www.wired.com/medtech/drugs/news/2005/12/69595 (accessed 12 July 2007).

Carsten, J. (2004) *After Kinship*. Cambridge: Cambridge University Press.

Cherian, J. (2006) 'India to emerge as global player in stem cell research', *All Headline News*, 18 April. Available at: www.allheadlinenews.com/articles/7003223259 (accessed 9 May 2006).

CII (2003) *Biotech India*. Available At: http://www.biotech-india.com

Clifford, J. (1997) *Routes: Travel and Translation in the Late Twentieth Century*. Cambridge, MA: Harvard University Press.

Clifford, J. and Marcus, G. E. (eds) (1986) *Writing Cultures: The Poetics and Politics of Ethnography*, Berkeley, CA: University of California Press.

Coghlan, A. (2003) 'Blood could generate body repair kit', *New Scientist*, 26 November.

—— (2004) 'Do you believe in miracles?', *New Scientist*, 9 October, pp. 36–9.

Cohen, L. (1999) 'Where it hurts: Indian materials for an ethics of organ transplantation', *Daedalus*, 128(4): 135–65.

—— (2001) 'The other kidney: biopolitics beyond recognition', *Body and Society*, 7 (2–3): 9–29.

—— (2004) 'Operability: surgery at the margin of the state', in V. Das and D. Poole (eds) *Anthropology in the Margins of the State*. Santa Fe: SAR Press, pp. 3–34.

—— (2005) 'Operability, bioavailability, and exception', in A. Ong and S. J. Collier (eds) *Global Assemblages: Technology, Politics, and Ethics as Anthropological Problems*. Oxford: Blackwell Publishing.

Collins, H. M. (1985) *Changing Order: Replication and Induction in Scientific Practice*. London: Sage.

Crabtree, P. (2006) 'Good old days gone for biotech', *The San Diego Union-Tribune*, 26 November.

Cyranoski, D. (2005) 'Consenting adults? Not necessarily … ', *Nature*, 435, 12 May, pp. 138–9.

Daily Telegraph (2007) 'Paraplegic walks after "stem cell" fix', 13 April.

Dalrymple, W. (2007) 'Business as usual: India's rise isn't a miraculous novelty, it's a return to traditional global trade patterns', *TIME*, August 13, p. 56.

Das, V. (1990) 'What do we mean by health?' In J. Caldwell and S. Fendley (eds) *In Health Transition*, Vol. II. Canberra: Australia National University.

—— (1995) 'Suffering, legitimacy and healing: the Bhopal case', in *Critical Events: An Anthropological Perspective on Contemporary India*, Delhi: Oxford University Press, pp. 137–74.

—— (2000) 'The practice of organ transplants: networks, documents, translations', in M. Lock *et al.* (eds) *Living and Working with the New Medical Technologies.* Cambridge: Cambridge University Press.

—— (2001) 'Stigma, contagion, defect: issues in the anthropology of public health', paper presented at Stigma and Global Health Conference. Bethesda, MD.

—— (2004) 'Signature of the state: the paradox of illegibility', in V. Das and D. Poole (eds) *Anthropology in the Margins of the State.* Santa Fe: School of American Research Press.

Das, V. and Addlakha, R. (2007) 'Disability and domestic citizenship: voice, gender and the making of the subject', in B. Ingstad and S. R. Whyte (eds) *Disability in Local and Global Worlds.* Berkeley, CA: University of California Press, pp. 128–48.

Das, V. and Poole, D. (2004) 'State and its margins: comparative ethnographies', in V. Das and D. Poole (eds) *Anthropology in the Margins of the State.* Santa Fe: SAR Press.

Dhawan, J., Gokhale, R. J. and Inder, I. M. (2005) 'Bioscience in India: times are changing', *Cell*, 123, December, 743–5.

Dhruvaranjan, V. (1989) *Hindu Women and the Power of Ideology.* Gramby, MA: Bergin and Garvey.

D'Monte, D. (2000) 'Ghandi's disputed heritage – protest in India against genetically modified crops', *UNESCO Courier*, January.

Douglas, M. (2005) *Purity and Danger: An Analysis of Concept of Pollution and Taboo.* London: Routledge.

Doyal, L. (2004) 'Informed consent: don't throw out the moral baby with the critical bathwater', *Quality and Safety in Health Care*, 13: 414–15.

Dugh, H. (2005) 'India threatens world order', *Express India*, Tuesday, 8 March.

Dyck, J. V. (1995) *Manufacturing Babies and Public Consent: Debating the New Reproductive Technologies.* London: Macmillan.

Edelman, M. (2005) 'Bringing the moral economy back into the study of 21st-century transnational peasant movements', *American Anthropologist*, 107(3): 331–45.

Eriksson, L. (2004) 'When scientists fight', *Science and Public Affairs*, June, 25.

Express Healthcare Management (2000) 'ICMR's proposed prohibition creates furore', available at: www.expresshealthcaremgmt.com

Fahim, H. (ed.) (1982) *Indigenous Anthropology in Non-Western Countries.* Durham, NC: Carolina Academic Press.

Farmer, P. (2003) *Pathologies of Power: Health, Human Rights, and the New War on the Poor.* Berkeley, CA: University of California Press.

Financial Express (2005) 'Firms with good track record to get stem cell R&D aid.', Saturday, 5 November. Available at: www.financialexpress.com

Fischer, M. J. (2003) *Emergent Forms of Life and the Anthropological Voice.* Durham, NC: Duke University Press.

Fochsen, G., Deshpande, K. and Thorson, A. (2006) 'Power imbalance and consumerism in the doctor–patient relationship: health care providers' experiences of patient encounters in a rural district of India', *Qualitative Health Research*, 16(9): 1236–51.

Franklin, S. (1990) 'Deconstructing "desperateness": the social construction of infertility in popular representations of new reproductive technologies', in P. McNeil *et al.* (eds) *The New Reproductive* Technologies. London: Macmillan.

—— (1997) *Embodied Progress: A Cultural Account of Assisted Conception.* London: Routledge.

—— (2001) 'Culturing biology: cell lines for the new millennium', *Health*, 5: 335–54.

—— (2003a) 'Ethical biocapital: new strategies of cell culture', in S. Franklin and M. Lock (eds) *Remaking Life and Death: Towards an Anthropology of the Biosciences.* Santa Fe: School of American Research Press.

—— (2003b) 'Kinship, genes, and cloning: life after Dolly', in D. Heath Goodman and M. S. Lindee (eds) *Genetic Nature/Culture: Anthropology and Science Beyond the Two-Culture Divide.* Berkeley, CA: University of California Press.

—— (2005) 'Stem Cells R Us: emerging life forms and the global biological', in A. Ong and S. J. Collier (eds) *Global Assemblages: Technology, Politics and Ethics as Anthropological Problems.* Oxford: Blackwell.

Frawley, D. (2001) 'The myth of the Hindu right', *Hinduism and the Clash of Civilizations,* Voice of India. Nov. Available at: http://www.vedanet.com/Hinduright.htm

Frew, S.E. *et al.* (2007) 'India's health biotech sector at a crossroads', *Nature Biotechnology* 25(4): 403–17.

Frontline (2001) 'At the forefront in India', *Frontline,* 15–28 September.

Fukuyama, F. (2003) *Our Posthuman Future: Consequences of the Biotechnology Revolution.* London: Profile Books.

Ganchoff, C. (2004) 'Regenerating movements: embryonic stem cells and the politics of potentiality', *Sociology of Health and Illness* 26(6): 757–74.

Geertz, C. (1983) 'Common sense as a cultural system', in C. Geertz, *Local Knowledge: Further Essays in Interpretive Anthropology.* New York: Basic Books.

Gentleman, A. (2007) 'Doubts over gene "cures"', *The Observer,* 3 June, p. 33.

Gibbon, S. and Novas, C. (eds) (2008) *Biosociality, Genetics and the Social Sciences: Making Biologies and Identities.* Abingdon: Routledge.

Ginsburg, F. D. and Rapp, R. (1995) *Conceiving the New World Order: The Global Politics of Reproduction.* Berkeley, CA: University of California Press.

Glasner, P. (2005) 'Banking on immortality? Exploring the stem cell supply chain from embryo to therapeutic application', *Current Sociology,* 53(2): 355–66.

—— (2007) 'Cowboy cloners, mavericks and kings: a cautionary tale of a promissory science', *21st Century Society: Journal of the Academy of Social Sciences,* 2(3): 265–74.

Glasner, P. and Rothman, H. (2004) *Splicing Life? The New Genetics and Society.* Aldershot: Ashgate.

Good, B. J. (1997) *Medicine, Rationality and Experience: An Anthropological Perspective.* Cambridge: Cambridge University Press

Gottweis, H. and Triendl, R. (2006) 'South Korean policy failure and the Hwang debacle', *Nature Biotechnology,* 24(2): 141–3.

Green, E., Griffiths, F., Henwood, F. and Wyatt, S. (2006) 'Desperately seeking certainty: bone densitometry, the internet and health care contexts', in A. Webster (ed.) *New Technologies in Health Care.* Basingstoke: Palgrave Macmillan.

Guardian (2006) 'Maverick medic reveals details of baby cloning experiment', *The Guardian* 20 July, p. 17.

Haran, J. (2007) 'Managing the boundaries between maverick cloners and mainstream scientists: the life cycle of a news event in a contested field', *New Genetics and Society,* in press.

Haran, J., Kitzinger, J., McNeil, M. and O'Riordan, K. (2007) *Human Cloning in the Media.* London: Routledge.

Haraway, D. (1990) 'A manifesto for cyborgs: science, technology, and socialist feminism in the 1980s', in L. J. Nicholson (ed.) *Feminism/Postmodernism.* London: Routledge.

—— (1991) *Simians, Cyborgs and Women: The Reinvention of Nature*. New York: Routledge.

Harding, L. (2003) 'India limbers up for space race as prime minister asks for the moon: shuttle astronaut inspires New Delhi to challenge China's lofty ambitions', *The Guardian*, 30 January.

Harlan, L. and Courtright, P. B. (eds) (1995) *From the Margins of Hindu Marriage: Essays on Gender, Religion and Culture*. New York: Oxford University Press.

Harvey, O. (2005) 'Regulating stem cell research and human cloning in an Australian context: an exercise in protecting the status of the human subject', *New Genetics and Society*, 24(2): 125–35.

Hedgecoe, A. and Martin, P. (2003) 'The drugs don't work: expectations and the shaping of pharmacogenetics', *Social Studies of Science*, 33: 327–64.

Hoeyer, K. (2003) '"Science is really needed – that's all I know": informed consent and the non-verbal practices of collecting blood for genetic research in northern Sweden', *New Genetics and Society*, 22(3): 229–44.

House of Lords (2002) *Stem Cell Research – Report*, House of Lords Stem Cell Research Committee, London: The Stationery Office.

Hundt, G. L., Green, J., Sandall, J., Hirst, J., Ahmed, S. and Hewison, J. (2006) 'Navigating the troubled waters of prenatal testing decisions', in A. Webster (ed.) *New Technologies in Health Care*. Basingstoke: Palgrave Macmillan.

ICMR (2000) *Ethical Guidelines for Biomedical Research on Human Subjects*. New Delhi: Indian Council for Medical Research.

—— (2004) *National Guidelines for Accreditation, Supervision and Regulation of ART Clinics in India*. New Delhi: Indian Council for Medical Research.

—— (2006) *ICMR-DBT Guidelines for Stem Cell Research and Therapy*. New Delhi: Indian Council for Medical Research and Department of Biotechnology.

Inhorn, M. (1996) *Infertility and Patriarchy: The Cultural Politics of Gender and Family Life in Egypt*. Philadelphia, PA: University of Pennsylvania Press

—— (2003) *Making Muslim Babies: Gender, Science, and In Vitro Technologies in Egypt*. New York: Routledge.

Inhorn, M. and Bharadwaj, A. (2007) 'Reproductively disabled lives: infertility, stigma and suffering in Egypt and India', in B. Ingstad and S. R. Whyte (eds) *Disability in Local and Global Worlds*. Berkeley, CA: University of California Press, pp. 78–106.

International Herald Tribune (2006) 'Australia lifts ban on cloning for stem cell research', *International Herald Tribune*, 6 December. Available at: www.iht.com (accessed 8 December 2006).

Jasanoff, S. (2005) *Designs on Nature: Science and Democracy in Europe and the United States*, Princeton, NJ: Princeton University Press.

—— (2006) 'Science policy for the public: civic epistemology and civic engagement', paper given at the 40th Anniversary Meeting of SPRU, 12 September, Science Policy Research Unit, University of Sussex.

Jayachandran, C. R. (2003) 'Indian ovum finds world market', *The Economic Times* 26, August.

Jayaraman, K. S. (2001) 'India seeks to block trade in human embryos', *Science and Development Network*, 27 September. Available at: http://www.scidev.net/

—— (2005a) 'Indian regulations fail to monitor growing stem-cell use in clinics', *Nature*, 434, 17 March, p. 259.

—— (2005b) 'Biotech boom', *Nature* 436: 480–1.

—— (2005c) 'Among the best', *Nature* 436: 492–5.

Kitzinger, J. and Williams, C. (2005) 'Forecasting science futures: legitimising hope and calming fears in the embryo stem cell debate', *Social Science and Medicine*, 61: 731–40.

Kitzinger, J., Williams, C. and Henderson, L. (2007) 'Science, media and society: the framing of bioethical debates around embryonic stem cell research between 2000 and 2005', in P. Glasner, P. Atkinson and H. Greenslade (eds) *New Genetics, New Social Formations*. London: Routledge, pp. 204–30.

Kleinman, A. (1995) *Writing at the Margin: Discourse Between Anthropology and Medicine*. Berkeley, CA: University of California Press.

Kumar, A.R. (2003) *Ethics of Human Cloning*. Keralakumudi. Available at: http://www.kaumudiusa.com/kusa/html/science_1.stm

Latour, B. (1993) *We Have Never Been Modern*. Cambridge, MA: Harvard University Press.

Lazaro, F. D. S. (2002) 'Stem cell research', *Online News Hour*. Public Broadcasting Service, 8 April. Available at: www.pbs.org

Lee, M. (2004) 'No "voodoo magic". But maybe a miracle cure', *The Independent on Sunday*, 31 October, p. 19.

Lévi-Strauss, C. (1969) *Elementary Structures of Kinship*. Boston: Beacon Press.

LifeSiteNews.com (2001) 'India forges ahead with embryo stem cell research', available at: http://www.lifesite.net

Lock, M. (2001) 'The alienation of body tissue and the biopolitics of immortalized cell lines', *Body & Society*, 7: 63–91

Lockhart Review (2005) *Legislative Review: Prohibition of Human Cloning Act 2002, and Research Involving Human Embryos Act 2002*, Legislative Review Committee (Chair Hon. John Lockhart). Canberra: Commonwealth of Australia.

Luce, E. (2006) 'One land, two planets', *New Statesman*, 30 January.

Marcus, G. E. (1995) 'Ethnography in/of the world system: the emergence of multi-sited ethnography', *Annual Review of Anthropology*, 24: 95–117.

Martin, E. (1994) *Flexible Bodies: The Role of Immunity in American Culture from the Days of Polio to the Age of AIDS*. Boston: Beacon Press.

Martin, P. (2001) 'Genetic governance: the risks, oversight and regulation of genetics databases in the UK', *New Genetics and Society*, 20(2): 157–83.

Marx, K. (1963) *Capital*, Vol. 1. Moscow: Progress Publishers.

Modell, J. (1994) *Kinship with Strangers: Adoption and Interpretation of Kinship in American Culture*. Berkeley, CA: University of California Press.

Mudur, G. (2004) 'Hospitals in India woo foreign patients', *British Medical Journal*, 328, 5 June, p. 1338.

Mudur, M. (2005) 'India plans to audit clinical trials', *British Medical Journal*, 331, p. 1044.

Müller, F. M. (ed.) (1894) *The Sacred Books of the East: Translations by Various Oriental Scholars and Edited by F. Max Müller*, Vols I–XLIX. Oxford: Clarendon Press.

Murdock, G. (2004) 'Popular representation and postnormal science: the struggle over genetically modified foods', in S. Braman (ed.) *Biotechnology and Communication. The Meta-Technologies of Information*, Hillsdale, NJ: Lawrence Erlbaum Associates, pp. 227–59.

Nandy, A. (1995) *Alternative Science: Creativity and Authenticity in Two Indian Scientists*. Delhi: Oxford University Press.

—— (1996) *Science, Hegemony and Violence: A Requiem for Modernity*. Delhi: Oxford University Press.

Napier, D. (2004) 'Emergent Forms of Life and the Anthropological Voice: Book Review', *Anthropological Quarterly*, 77(4): 859–62.

Narayan, K. (1993) 'How native is a native anthropologist?' *American Anthropologist*, 95: 19–34.

Nature (2005) 'Nature Outlook: India', *Nature* 436: 477–98.

—— (2006) 'Come veto or high water', Editorial, *Nature*, 442: 329.

New Scientist (2005) 'The next knowledge superpower', *New Scientist*, 185, 19 February, pp. 31–2.

Nichter, M. and Nichter, M. (1996) *Anthropology and International Health: Asian Case Studies*, 2nd edn. Amsterdam: Gordan and Breach.

Nuffield (2000) *Stem Cell Therapy: The Ethical Issues*. London: The Nuffield Council on Bioethics.

Ohnuki-Tierney, E. (1984) '"Native" anthropologists', *American Ethnologist*, 1(3): 584–6.

Ong, A. (2006) *Neoliberalism as Exception: Mutations in Citizenship and Sovereignty.* Durham, NC: Duke University Press.

Ong, A. and Collier, S. (eds) (2005) *Global Assemblages: Technology, Politics and Ethics as Anthropological Problems.* Oxford: Blackwell.

Pandeya, R. (2007) 'Celling the elixir of life', *Business Standard*, 6 May. Available at: www.business-standard.com/general/printpage.php?autono = 283528 (accessed 18 May 2005).

Patel, T. (1994) *Fertility Behaviour: Population and Society in a Rajasthan Village.* Delhi: Oxford University Press.

Pearson, C. (2006) 'Sold on misconceptions', *Weekend Australian*, 19–20 August, p. 30.

Petersen, A. (2002) 'Replicating our bodies, losing our selves: news media portrayals of human cloning in the wake of Dolly', *Body and Society*, 8(4): 71–90.

Petryna, A. (2005) 'Ethical variability: drug development and globalizing clinical trials', *American Ethnologist*, 32(2): 183–97.

Pilnick, A. (2002) *Genetics and Society. An Introduction.* Buckingham: Open University Press.

Prakash, G. (1999) *Another Reason: Science and the Imagination of Modern India.* Princeton, NJ: Princeton University Press.

Prasad, M. G. (2001) *Cellular Division, Science and Spirit.* Available at: http://www.science-spirit.org

Rabinow, P. (1992) 'Artificiality and enlightenment: from sociobiology to biosociality', in J. Crary and S. Kwinter (eds) *Incorporations.* New York: Zone.

—— (1996) *Essays on the Anthropology of Reason.* Princeton, NJ: Princeton University Press.

—— (1999) *French DNA: Trouble in Purgatory.* Chicago: The University of Chicago Press.

Rabinow, P. and Dan-Cohen, T. (2005) *A Machine to Make a Future: Biotech Chronicles.* Princeton, NJ: Princeton University Press.

Radhakrishnan, S. (1978) *The Principal Upanishads.* New Delhi: Indus.

Rajan, K. S. (2006) *Biocapital. The Constitution of Postgenomic Life.* Durham, NC: Duke University Press.

Randerson, J. (2005a) 'Vaccines for pennies: they said it couldn't be done', *New Scientist*, 185, 19 February, p. 42.

—— (2005b) 'Sight for sore eyes: an Indian hospital is pioneering a stem-cell cure for blindness', *New Scientist*, 185, 19 February, pp. 50–1.

Rapp, R. (2000) 'Extra chromosomes and blue tulips: medico-familial interpretations', in M. Lock, A. Young and A. Cambrosio (eds) *Living and Working with the New Medical Technologies: Intersections of Inquiry.* Cambridge: Cambridge University Press.

Rediff.com. (2004) 'Clinical trials: why India is irresistible', 22 December. Available at: http://www.rediff.com/money/2004/dec/22spec.htm

Relman, A. (1987) 'Sounding board: the changing climate of medical practice', *New England Journal of Medicine*, 31696): 333–42.

Riesman, C. (2000) 'Stigma and everyday resistance practices: childless women in South India', *Gender and Society*, 14(1): 111–35.

Roberts, E.F.S. (2008) 'Biology, sociality and reproductive modernity in Ecuadorian in-vitro fertilization: the particulars of place', in S. Gibbon and C. Novas (eds) *Genetics, Biosociality and the Social Sciences: Making Biologies and Identities.* London: Routledge, pp. 79–97.

Rose, H. (2001) 'Gendered genetics in Iceland', *New Genetics and Society*, 20(2): 119–38.

—— (2004) 'Beware the cowboy cloners', *The Guardian*, 16 February, p. 16.

Rose, N. and Novas, C. (2005) 'Biological citizenship', in A. Ong and S. J. Collier (eds) *Global Assemblages: Technology, Politics, and Ethics as Anthropological Problems.* Oxford: Blackwell.

Salter, B., Cooper, M. and Dickens, A. (2006) 'China and the global stem cell bioeconomy: an emerging political political strategy?', *Regenerative Medicine*, 1(5): 671–83.

Salter, B., Cooper, M., Dickens, A. and Cardo, V. (2007) 'Stem cell science in India: emerging economies and the politics of globalization', *Regenerative Medicine*, 2(1): 75–89.

Sample, I. (2006) 'Stem Cell Bank to begin supplying researchers', *The Guardian*, 18 September, p. 9.

Sardar, Z. (2004) 'Introduction: the A, B, C, D (and E) of Ashis Nandy', in *Return from Exile: Alternative Science, The Illegitimacy of Nationalism, The Savage Freud.* Delhi: Oxford University Press.

Scheper-Hughes, N. (1992) *Death Without Weeping: The Violence of Everyday Life in Brazil.* Berkeley, CA: University of California Press.

—— (2000) 'The global traffic in human organs', *Current Anthropology*, 41(2): 191–224.

—— (2001) 'Bodies for sale—whole or in parts', *Body & Society*: Special Issue on Commodifying Bodies, ed. N. Scheper-Hughes and L. Waquant, 1–8: 31–62.

Schwartz, R.S. (2006) 'The politics and promise of stem-cell research', *The New England Journal of Medicine*, 355(12): 1189–91.

Scott, J. (1976) *The Moral Economy of the Peasant: Rebellion and Subsistence in Southeast Asia.* New Haven, CT: Yale University Press.

—— (1985) *Weapons of the Weak: Everyday Forms of Peasant Resistance.* New Haven, CT: Yale University Press.

—— (2005) 'Afterword to "Moral Economies, State Spaces and Categorical Violence"', *American Anthropologist,* 107(3): 395–402.

Sengupta, A. and Nundy, S. (2005) 'The private health care sector in India', *British Medical Journal*, 331, 19 November, pp. 1157–8.

Sharma, A. (2006) 'Stem cell research in India: emerging scenario and policy concerns', *Asian Biotechnology and Development Review*, 8(3): 43–53.

Sigurdsson, S. (2001) 'Ying-yang genetics, or the HSD DeCode controversy', *New Genetics and Society*, 20(2): 102–17.

Sky News (2006) 'The Stem Cell "Miracles"', Monday, 23 January. Available at: http://news.sky.com/skynews/article/0,30000-1209738,00.html

Sky News (2007) 'Miracle Stem Cell Cure?' Friday, 13 April. Available at: http://news.sky.com/skynews/article/0,30200-1260433,00.html

Soja, E. W. (1996) *Thirdspace: Journeys to Los Angeles and Other Real-and-Imagined Places*. Cambridge, MA: Blackwell.

Srinivas, M. N. (1997) 'Practicing social anthropology in India', *Annual Review of Anthropology*, 26: 1–24.

Srinivasan, S. (2006) 'Clinical trials – Part IV: Rogue research in the guise of stem cell therapy', infochangeindia.org, March. Available at: www.infochangeindia.org/features326.jsp (accessed 18 May 2007).

Stark, R. (1985) 'Lay workers in primary health care: victims in the process of social transformation', *Social Science and Medicine*, 20: 269–75.

Stephens, N., Atkinson, P. and Glasner, P. (2007) 'Bridging strategies and husbandry in the UK Stem Cell Bank: closing the regulators' regress', mimeo, CESAGEN, Cardiff University (submitted for publication).

—— (2008) 'Securing the past and validating the present to protect the future: regulation at the UK Stem Cell Bank', *Science as Culture*, 17(1): 43–56.

Strathern, M. (1988) *The Gender of the Gift: Problems with Women and Problems with Society in Melanesia*. Berkeley, CA: University of California Press.

—— (1992) *Reproducing the Future: Anthropology, Kinship and the New Reproductive Technologies*. New York: Routledge.

Subramaniam, B. (2000) 'Archaic modernities: science, secularism and religion in modern India', *Social Text*, 18(3): 67–86.

—— (2002) 'Colonial legacies and the postcolonial predicament: the case of modern India', paper presented at the Society for the Social Studies of Science Conference, Milwaukee.

Sunder Rajan, K. (2002) 'Banking (on) biologicals: commodifying the global circulation of human genetic material', *Sarai Reader*, 277–89.

—— (2006) *Biocapital: The Constitution of Postgenomic Life*. Durham, NC: Duke University Press.

—— (2008) 'Biocapital as an emergent form of life: speculations on the figure of the experimental subject', in S. Gibbon and C. Novas (eds) *Biosociality, Genetics and the Social Sciences: Making Biologies and Identities*. London: Routledge, pp. 157–87.

Thapar-Björkert, S. (1999) 'Negotiating otherness: dilemmas for a non-Western researcher in the Indian sub-continent', *Journal of Gender Studies*, 8(1): 57–69.

The Hindu (2007) 'Guidelines being framed to regulate stem cell research', *The Hindu*, 30 January. Available at: www.thehindu.com/2007/01/30/stories/2007013004440400.htm (accessed 26/07/2007).

The Indian Express (2005) 'Stem cells: "Miracles" in the dark', 3 April.

The Week (2001) 'The stem cell saga: a lost science of India?' 16 September.

Thompson, C. (2005) *Making Parents: The Ontological Choreography of Reproductive Technologies*. Cambridge, MA: MIT Press.

—— (2001) 'Strategic naturalizing: kinship in an infertility clinic,' in S. Franklin and S. McKinnon (eds) *Relative Values: Reconfiguring Kinship Studies*. Durham, NC: Duke University Press.

Thompson, E. P. (1971) 'The moral economy of the English crowd in the eighteenth century', *Past and Present* 50: 76–136.

Thorold, C. (2001) 'Indian firms embrace biotechnology,' BBC News, 6 April. Available at: http://news.bbc.co.uk/1/hi/world/south_asia/1264569.stm

Throop, J. C. (2005) 'Hypocognition, a "sense of the uncanny," and the anthropology of ambiguity: reflections on Robert I. Levy's contribution to theories of experience in anthropology', *Ethos*, 33(4): 499–511.

TIME (2007) India charges ahead: sixty years of independence , 13 August.

Times India (2005a) 'AIIMS claims cutting edge stem cell study', *The Times of India*, 23 March.

—— (2005b) 'Celling success', editorial in *The Times of India*, 10 March.

Timmermans, S. and Berg, M. (2003) *The Gold Standard: The Challenge of Evidence-Based Medicine and Standardization in Health Care*. Philadelphia, PA: Temple University Press.

Turner, V. (1967) *The Forest of Symbols: Aspects of Ndembu Ritual*. Ithaca, NY: Cornell University Press.

—— (1969) *The Ritual Process: Structure and Anti-structure*. Chicago: Aldine Publishing Co.

—— (1974) *Dramas, Fields and Metaphors: Symbolic Action in Human Society*. Ithaca, NY: Cornell University Press.

—— (1975) *Revelation and Divination in Ndembu Ritual*. Ithaca, NY: Cornell University Press.

Turney, J. (1998) *Frankenstein's Footsteps: Science, Genetics and Popular Culture*, New Haven, CT: Yale University Press.

Tyagananda, S. (2002) *Stem Cell Research: A Hindu Perspective*. Available at: http://home.earthlink.net/~tyag/Home.htm

Uberoi, P. (2006) *Freedom and Destiny: Gender, Family and Popular Culture in India*. New Delhi: Oxford University Press.

UNICEF (2006) Available at: http://www.unicef.org/

Van Gennep, A. (1960) *The Rites of Passage*. London: Routledge.

Van Hollen, C. (2003) *Birth on the Threshold: Childbirth and Modernity in South India*. Berkeley, CA: University of California Press.

Waldby, C. (2002) 'Stem cells, tissue cultures and the production of biovalue', *Health*, 6: 305–23.

—— (2003) 'The UK Stem Cell Bank: managing the tissue economy', paper given to the ESRC Stem Cell Research Workshop on Developing the Economic and Social Agenda, 17 June, Royal College of Obstetricians and Gynaecologists, London.

Waldby, C. and Mitchell, R. (2006) *Tissue Economies: Blood, Organs and Cell Lines in Late Capitalism*. Durham, NC: Duke University Press.

Walsh, N. P. (2006) 'India flexes its muscles with first foreign military base', *The Guardian*, April 26.

Washington Post (2001) 'India plans to fill void in stem cell research: scientists say restrictions in U.S. may give them advantage in development', 28 August, A07.

Webb, J. (2005) 'Mashelkar, R.A. is campaigning for a new way of thinking', *New Scientist*, 185, 19 February, p. 41.

Wellcome Trust (1998) *Public Perspectives on Human Cloning*. London: The Wellcome Trust.

Whyte, S. R. and Ingstad, B. (2007) 'Introduction: disability connections', in B. Ingstad and S. R. Whyte (eds) *Disability in Local and Global Worlds*. Berkeley, CA: University of California Press, pp. 1–29.

Wilson, C. (2005) 'Voyage of discovery: India's drug companies are risking everything to stay afloat', *New Scientist*, 185, 19 February, pp. 44–7.

Winickoff, D. E. (2006) 'Governing stem cell research in California and the USA: towards a social infrastructure', *TRENDS in Biotechnology*, 24: 9390–4.

Wittgenstein, L. (1953) *Philosophical Investigations.* Oxford: Blackwell.

Wood, A. (1996a) *Interpreting the Upanishads.* Pune: Ananda Wood.

—— (1996b) *From the Upanishads.* Pune: Ananda Wood.

World Bank (2006) Available at: http://www.worldbank.org/

Zilinkas, R. A. (1995) 'Biotechnology and the Third World: the missing link between research and applications', in M. Fransman *et al.* (eds), *The Biotechnology Revolution?* Oxford: Blackwell.

Index

Milton Keynes UK
Ingram Content Group UK Ltd.
UKHW031531071024
449327UK00005B/136

9 780415 396097